CONSTRUCTING REFORM: ECONOMIC EXPERTISE AND THE GOVERNING OF AGRICULTURAL CHANGE IN AUSTRALIA

CONSTRUCTING REFORM: ECONOMIC EXPERTISE AND THE GOVERNING OF AGRICULTURAL CHANGE IN AUSTRALIA

VAUGHAN J. G. HIGGINS

Nova Science Publishers, Inc.
New York

Coordinating Editor: Tatiana Shohov
Senior Editors: Susan Boriotti and Donna Dennis
Office Manager: Annette Hellinger
Graphics: Wanda Serrano
Editorial Production: Vladimir Klestov, Matthew Kozlowski and Maya Columbus
Circulation: Ave Maria Gonzalez, Vera Popovich, Luis Aviles, Raymond Davis,
Melissa Diaz and Jeannie Pappas
Communications and Acquisitions: Serge P. Shohov
Marketing: Cathy DeGregory

Library of Congress Cataloging-in-Publication Data
Available Upon Request

ISBN 1-59033-604-6.

400 Oser Ave, Suite 1600
Hauppauge, New York 11788-3619
Tele. 631-231-7269 Fax 631-231-8175
e-mail: Novascience@earthlink.net
Web Site: http://www.novapublishers.com

Printed in the United States of America

CONTENTS

LIST OF TABLES

LIST OF ABBREVIATIONS

ABARE	Australian Bureau of Agricultural and Resource Economics
ABS	Australian Bureau of Statistics
ARMCANZ	Agricultural and Resource Management Council of Australia and New Zealand
AMIEU	Australian Meat Industries Employment Union
BAE	Bureau of Agricultural Economics
CUA	Cattlemen's Union of Australia
DPI	Australian Department of Primary Industry
DPIE	Australian Department of Primary Industries and Energy
DRP	Drought Relief Payment
EC	Exceptional Circumstances
FarmBis	Farm Business Improvement Program
FMB	Farm Management Bonds
FMD	Farm Management Deposits Scheme
FUW	Farmers' Union of Wales
IAC	Industries Assistance Commission
IED	Income Equalization Deposits
IMF	International Monetary Fund
NFF	National Farmers' Federation
NFU	National Farmers' Union
PIPES	Queensland Primary Industry Productivity Improvement Scheme
PMP	Property Management Planning
QDPI	Queensland Department of Primary Industries
QIDC	Queensland Industry Development Corporation
QRAA	Queensland Rural Adjustment Authority
RAS	Rural Adjustment Scheme
RASAC	Rural Adjustment Scheme Advisory Council
RIRDC	Rural Industries Research and Development Corporation
RRS	Rural Reconstruction Scheme
Triple A	Agriculture – Advancing Australia Policy Package: released 1997
UGA	United Graziers Association

ACKNOWLEDGMENTS

This book would not have been possible without the generous assistance and support of numerous people. First and foremost I would like to thank my PhD supervisors, Professor Geoffrey Lawrence and Associate Professor Daniela Stehlik, for their endless encouragement, their generosity in sharing valuable knowledge and skills, and their willingness to read numerous chapter drafts. Without their immense assistance and advice, this monograph would not have been possible.

I would like also to express my sincere thanks to the participants whom I interviewed in the course of this study, and who willingly took time out of their busy schedules to speak with me. Their views made an invaluable contribution to this project and I thank them for their participation.

Numerous people also provided specific assistance and encouragement throughout this project including Stewart Lockie, Simon Kitto, Morgan Brigg, Lynda Herbert-Cheshire, Kristen Lyons, Bruce Taylor, Jane Muller, Harry Ballis, Steven Russell, Marion Collis and Kerryn O'Halloran.

Last, but certainly not least, I acknowledge the continued support of my direct and extended family in both Australia and New Zealand.

Earlier versions and parts of the following chapters have been published elsewhere.

Chapter 1 (2001) in *Sociologia Ruralis*, 41(3)
Chapter 3 (2001) in *Review of International Political Economy*, 8(2), http://www.tandf.co.uk
Chapters 4 and 5 (2001) in *Journal of Sociology*, 37(3), and (2002) *ARENA journal*

INTRODUCTION – A FRAMEWORK FOR STUDYING THE GOVERNANCE OF AGRICULTURAL CHANGE

The productivity and efficiency of farmers in both Australia and other Western nations has assumed increasing significance on the formal political agenda since the 1960s. Economic changes at both the international and national level have raised questions among economists, farm organizations and state agencies concerning the capacity of some farmers to be able to earn an adequate income, and thereby contribute to national agricultural output and efficiency. While these concerns have contributed to policies of farm subsidization in Europe, Britain and the United States, Australia has taken a somewhat different path. Farm reconstruction and adjustment programs were created from the early-1970s that sought to 'restructure' the farm sector by encouraging so-called 'unviable' farmers to exit the industry while providing incentives for remaining producers to improve their productivity.

Economists in both the public service and academic institutions have welcomed such policies arguing that intervention in assisting processes of 'adjustment' makes agricultural resources available for those farmers who will use them in the most efficient manner, while ensuring also that the welfare needs of unviable farmers are met (eg. Standen and Musgrave, 1969; Harris, 1970; Bureau of Agricultural Economics, 1973, 1974; Kingma and Samuel, 1977; Musgrave, 1990; Gow, 1996; Malcolm, Davidson and Vandenburg, 2000). According to this perspective, restructuring of the farm sector is an inevitable consequence of market-driven processes that needs to be theorized and studied so that both governments and farmers can be informed on how to manage it efficiently. Agricultural economists are accorded a central role in 'discovering' the causes of viability problems, and in diagnosing solutions that smooth seemingly inevitable, and external, processes of change. The broad aim of this book is to explore the events, techniques and forms of calculation through which economic expertise in Australia has come to assume increased authority on matters of agricultural change, and the consequences of this for how strategies of 'restructuring' are assembled as objects of knowledge.

BACKGROUND TO THE STUDY

Exit from farming has been an ongoing feature of Australian agriculture since the commencement of European settlement (Musgrave, 1990; Burdon, 1996). However, state assistance to encourage some farmers to leave their land, while providing incentives for others to build up their properties, has only been provided consistently since the early-1970s. Policies such as the Marginal Dairy Farms Reconstruction Scheme (1970), and the more broadly-based Rural Reconstruction Scheme (1971), were established initially as a response to downturns in the dairy and wheat/wool industries. However, events such as the entry of Britain into the European Economic Community in 1973, and the OPEC oil shock, convinced many economists that these downturns were part of longer-term structural economic problems. The introduction of the Rural Adjustment Scheme (RAS) in 1977 represented an attempt by state agencies to smooth the process of change – constructed through the work of economists – and improve the fortunes of Australian agriculture, by providing long-term assistance measures to redistribute agricultural resources to those who would use them most efficiently.

From 1977 to 1997 the Rural Adjustment Scheme (RAS) was the main vehicle for regulating change in Australian agriculture. The objectives and specific programs of the RAS changed a number of times from its creation with key reviews in 1983/84 (Industries Assistance Commission, 1984), 1987 (Coopers and Lybrand WD Scott, 1988), 1992 (Synapse Consulting, 1992), and in 1996/97 (McColl, Donald and Shearer, 1997). A gradual shift is evident over this period from a desire to spread resources 'as widely as practicable' (see Commonwealth of Australia, 1976a), toward objectives that encouraged formal business skills as the best means of operating a viable farm and improving economic efficiency.

The adoption of formal business management skills as a means of risk management on the farm is argued in official reviews to result in less distorted market signals, and be both necessary and desirable in the interests of efficiency, productivity and sustainability (eg. Kerin and Cook, 1988; Synapse Consulting, 1992; Commonwealth of Australia, 1994, 1995, 1996; McColl et al., 1997). Such a view is shared by Australia's peak farming body, the National Farmers' Federation (NFF, 1993) and, with some qualifications, by a number of academic agricultural economists (see for example, Cook, Edwards and Ronan, 1994; Stayner, 1994; Gow, 1996; Malcolm et al., 2000). The encouragement of individual risk management (or 'self-reliance' to use the policy terminology) is now the central objective of many farm programs, and is argued in official documents to represent the most rational means of improving agricultural productivity and environmental sustainability. This is seen by some to indicate the gradual emergence of a coherent economic rationalist approach to governing the conduct of farmers (Cockfield, 1995; see also Pusey, 1991 for a more general analysis of the emergence of an 'economic rationalist' approach to policy-making).

In opposition to the view that successive farm adjustment schemes represent part of a progression toward more effective ways of managing agricultural change, sociologists have emphasised the broader relations of production within which these programs are embedded. Of the small amount of sociological research devoted specifically to this area, policies seeking to manage change are conceptualized as part of a neo-liberal approach to policy-making argued to represent part of a broader crisis of the state in which regulation is geared increasingly toward the profit-making interests of powerful transnational corporations (Lawrence, 1987, 1992; Higgins, 1999). Applied more generally to analyses of agricultural regulation, sociologists view the state as acting in accordance with the profit-making interests of transnational capital. As a consequence, states are compelled to abolish programs of social support in favor of policies that promote international competitiveness, deregulation, labour market reform and economic efficiency (McMichael and Myhre, 1991; Marsden, 1992; Le

Heron, 1993; Bonanno, Busch, Friedland, Gouveia and Mingione, 1994; Buttel, 1996; McMichael, 1999; Campbell and Lawrence, 2000).

The literature on agricultural restructuring accords with broader theories of capitalist regulation. While by no means a coherent body of literature, these theories generally claim that the regulatory capacities of the state to project its power are weakened by an intensification of global capital accumulation, new production systems, and the growing power of transnational corporations and global regulatory bodies (eg. Aglietta, 1987; Lipietz, 1987; Held, 1991; Lipietz, 1992; Jessop, 1994a, 1994b; Axford, 1995; McMichael, 1996; Strange, 1996; Jessop, 1997). A neo-liberal political agenda that focuses on competitiveness and efficiency is argued by many to be the ultimate effect of these changes in which economic management via a Keynesian welfare-state is displaced by a rationalist faith in market forces to solve policy problems (applied to Australia see Pusey, 1991; Bryson, 1992; Muetzelfeldt, 1992; Stilwell, 1993). In this sense, the state does not so much retreat from regulation, but maintains a significant role in the process of restructuring, enhancing the international competitiveness of industries and forming partnerships with non-state actors (Jessop, 1994a; Jessop, 1994b; Hirst and Thompson, 1996; Weiss, 1997). From this perspective, contemporary 'market-based' approaches to agricultural restructuring are part of a broad shift away from the welfare-state as a form of regulation.

Scope and Purpose of the Book

While the perspectives outlined above may be compelling in explaining why particular forms of regulation rather than others have been used to manage contemporary agricultural change, they tend not to examine how this has been possible. In representing social and economic change as a causal process underpinned by a structural logic, such as the pursuit of increased efficiency, or intensified capital accumulation, these perspectives overlook the diversity of objectives, forms of knowledge, and material techniques through which specific forms of agricultural regulation have assumed prominence as governable problems. What this means is that economists and many sociologists tend to focus exclusively on the macro-structuring of reality, without opening the *black boxes* that constitute these structures (see Callon and Latour, 1981).

This book seeks to challenge those explanations of history based on structural causes of change by examining *how* a field of concerns known as agricultural restructuring has become a thinkable and a governable issue. Specifically, it seeks to examine the authorities, forms of knowledge, and types of intervention that have characterized attempts to construct and govern a 'restructuring problem' in Australian agriculture. In this sense, the book focuses on the specific conditions of possibility that have enabled contemporary forms of thinking about and managing change to emerge as objects of governance. It argues, in a similar manner to Murdoch (1995), Martin (1997a, 1997b), Murdoch and Ward (1997) and Lockie (1999), for an understanding of the subtle discourses and techniques through which agricultural problems are constituted as knowable and governable and, in particular, the role of economic expertise in this process. Therefore, the book is not an attempt either to evaluate how the practices of farmers are really governed (if this is at all possible), or to conduct a policy analysis of programs seeking to restructure the Australian farm sector. Rather than searching for explanations concerning the 'real' causes and consequences of agricultural restructuring, the broad aim is to reconstruct the strategies pursued in Australia between 1967 and 1997 through which it has been constituted as both knowable and governable. This period in history is chosen for a particular reason. The year 1967 marked the emergence of a 'low-income farm problem' seen to have dire implications for agricultural productivity and the welfare of farmers (see McKay, 1967), while the year 1997 marked a major review of programs seeking to govern agricultural change. Significantly, the study of this 30 year time period provides an insight into the different ways in which 'restructuring' was reflected upon and governed as a problem that could truthfully be explained and managed only by economic expertise. In particular, it enables the contemporary policy focus on self-reliance and individual risk management to be analyzed as part of historical attempts to 'know' and to govern processes of agricultural change.

This book focuses particularly on the Commonwealth or Federal level of government in making restructuring knowable and governable in particular ways, although it examines also the role of the Australian States and farm organizations. Australia has three levels of government: Commonwealth, State and Local. Only the Commonwealth government has national jurisdiction. The Commonwealth, under Section 96 of the Federal Constitution, has historically taken responsibility for programs seeking to manage agricultural change, setting the main objectives and ensuring coordination among the States. Such coordination is necessary as the States (and from 1985 the Northern Territory) were given administrative responsibility for these programs. From the early 1970s, each State established an official authority responsible for administering adjustment/reconstruction programs, making their practices crucial in understanding how Commonwealth objectives are governed. Special reference will be made in this book to the State of Queensland as a specific example of the ways in which State authorities have sought to interpret Commonwealth objectives. Farm organizations are also significant in terms of the influence they have sought to have on the governing of restructuring, especially since the creation of the National Farmers' Federation (NFF) in 1979 (see Connors, 1996).

METHOD

Understanding how particular ways of knowing and governing agricultural change have been assembled historically, and the role of economic expertise in this process, requires analytical concepts and tools that move beyond structuralist economic and sociological accounts that tend to impute a causal dynamic to social change. This book draws upon a genealogical framework developed from the work of French philosopher Michel Foucault. A genealogical approach questions history as the search for hidden meaning, and the foundations or origins of knowledge. Foundationalism is the belief that knowledge can be securely grounded or validated so as to provide accurate access to reality (Yeatman, 1994). According to this view, (social) science is a quest for the true foundations of knowledge through which universal claims to truth can be established. It is those *meta-narratives* (Lyotard, 1984) that make explicit appeal to some final foundation of truth such as 'the hermeneutics of meaning, the emancipation of the rational or working subject, or the creation of wealth' (xxiii), against which genealogy wages its battle.

Genealogy seeks to 'dispel the chimera of the origin' (Foucault, 1984a:80), and show the arbitrariness and contingencies upon which the present is based. Knowledge is here conceptualized as part of a 'will to truth' (Foucault, 1984b), that attempts to 'filter, hierarchize and order' historical events 'in the name of some true knowledge and some arbitrary idea of what constitutes a science and its objects' (Foucault, 1980:83). Genealogy provides a conceptually coherent means for challenging the search in historical investigations for the origins or foundations of knowledge, by enabling the production of an *effective* history (see Dean, 1994) that records events without reference to an underlying signifier. It exposes the struggles for truth that underpin the claimed neutrality of these knowledges. This involves re-conceptualizing the status of knowledge from a means to secure rationality and social progress, to a site of political struggle in which the power to define and impose meaning takes place (Foucault, 1972). The role of knowledge in ordering, categorizing and giving meaning to social life means that it is intimately linked to power and the production of truth. From this perspective, there is no longer any underlying dynamic or fixed meanings to be discovered: history is a series of imposed interpretations. The task of genealogy is not to find a universal basis for understanding these interpretations, nor to investigate whether they are really telling the truth, but to record them in their singularity as a series of *events* without recourse to a unifying signifier or cause and 'outside of any monotonous finality' (Foucault, 1984a:76).

An *analytics of government* (see Miller and Rose, 1990; Rose and Miller, 1992; Rose, 1993; Dean, 1999; Rose, 1999) provides the theoretical basis for linking agricultural change as a governable problem to a genealogical historical framework. Drawing upon Foucault's writings on governmentality (see particularly Foucault, 1991), this approach focuses on how practices of governing are constituted in a knowable and manageable form, and particularly the rationalities and technologies through which these practices are constituted and changed. Rather than starting from a global theory of the state or regulation, an analytics of government seeks to examine 'how different locales are constituted as authoritative and powerful, how different agents are assembled with specific powers, and how different domains are constituted as governable and administrable' (Dean, 1999:29).

This style of governmentality studies pays particular attention to the relationship between rationalities or ways of reflecting on rule, and the technical means for giving effect to governmental ambitions. Political rationalities are concerned with the proper ends, means and limits of government. They are what make programs of government thinkable and knowable as objects of knowledge and include, according to Miller and Rose (1990:6), 'philosophical doctrines, notions of social and human realities, theories of power, conceptions of policy and versions of justice'. Drawing on the work of Foucault, rationalities can be conceptualised as more or less systematic discursive means for organising social life on the basis of knowledge deemed to be truthful. As Rose (1999:28) argues:

> Political rationalities are discursive fields characterized by a shared vocabulary within which disputes can be organized, by ethical principles that can communicate with one another, by mutually intelligible explanatory logics, by commonly accepted facts, by significant agreement on political problems.

These discursive fields, however, are not simply reflections of reality. They enable reality to be rendered thinkable, calculable, and therefore amenable to intervention, in a particular way. In other words, they enable government to be assembled in a programmatic and technically 'knowable' form.

Governing is not based simply on discursive rationalities of rule, through which 'reality' is ordered, categorized and made knowable, but on what Dean (1999:31) calls the technical means that enable rule to be practically possible. Rose and Miller (1992) emphasise that an analytics of governing should focus on more than simply rule as forms of rationality: reflections on rule cannot be adequately understood without examining political technologies – the 'complex of mundane programmes, calculations, techniques, apparatuses, documents and procedures through which authorities seek to embody and give effect to governmental ambitions' (175). Political technologies are a means for transforming rationalities of government into a technical means for shaping conduct 'at a distance' in that they allow reality to be made stable, mobile, comparable and combinable (see Latour, 1988). They can be conceptualized as complex assemblages of knowledge, expertise, calculation, representation and inscription that seek to shape conduct in particular directions for particular purposes (Dean, 1996). In this sense, technologies are not simply material but comprise also what Foucault (1988:18) calls a 'matrix of practical reason' that imply certain modes of training and modification of individuals. However, the ways in which these technologies are applied cannot be reduced to particular political rationalities. As Dean (1994:188) argues, 'political rationality may codify and assemble particular technologies within various programs, but the technologies themselves are a condition for that rationality and have forms of rationality inscribed within them'. This suggests that rationalities and technologies are relational and fluid. Significantly, technologies of rule do not simply reflect broader political rationalities but, once applied, may have unforseen or undesirable effects, shaping both how rule is reflected upon and the types of technologies seen to produce the most 'effective' outcomes.

Expertise is crucial in enabling political rationalities and technologies to be assembled in a way that shapes conduct and contributes to programs achieving their desired effects. In order for the conduct of individuals to be shaped according to political objectives, authorities must ensure that calculations and strategies formulated at one location can be linked to action another. Knowledge plays a significant role in connecting the objectives of political authorities to the actions of individuals in heterogeneous micro-locales as the 'expert' knowledges of agents such as doctors, managers, planners, economists and social workers transform the concerns of government into the technical and non-political language of truth (Rose, 1993). As Rose and Miller (1992:188) state, expertise translates the political concerns of government – efficiency, industrial productivity, law and order, normality – into the politically neutral discourse of management, social science, accounting and so forth.

Armed with rationalities and technologies that promise to produce 'truthful' outcomes, these forms of expertise seek to shape conduct in desirable directions by teaching individuals the ways in which they might manage their finances better, live in a healthier way or work more efficiently (Rose and Miller, 1992:188). In this way, individual preferences and choices are aligned with political objectives through the seemingly 'voluntary' enhancing of self-regulatory capacities, rather than by means of coercion. The apparently non-political activities and calculations of experts allow private matters to be addressed while maintaining the formal freedom of subjects.

Governmental rationalities, technologies and forms of expertise have been assembled in various ways in the last 100 years. While such assemblages are complex and contested, two specific 'regimes of government' can be identified within liberal democracies (Dean, 1999): *welfarism* and *advanced liberalism*. Briefly, welfarism involves the encouragement of national growth and well being through the promotion of social responsibility and the mutuality of risk, while advanced liberalism seeks to optimise economic growth through the entrepreneurship of autonomous actors (Rose, 1993). It is crucial to note that these regimes of governing should not be considered as distinct epochs or institutions (as in the 'welfare state' for example), but as ways of integrating various rationalities and technologies so that they appear 'to partake in a coherent logic' (Rose, 1999:27).

From the perspective of a genealogical analytics of government outlined above, attempts to restructure Australian agriculture is conceptualised in this book as an *effect* of the different rationalities, technologies and forms of expertise through which governing as a practical exercise is assembled. This suggests that different regimes of government embody particular ways of reflecting upon, and techniques for the governing of, agricultural change. In fact, one of the purposes of this book is to show how an advanced liberal regime of government has sought increasingly to constitute the proper management of change as an issue of individual risk management, rather than as a collective problem to be governed through welfarist forms of rule.

The methodological task of studying the governing of agricultural change using a genealogical analytics of government involves an examination of the rationalities of rule, authorities, forms of conduct, objectives and material techniques through which it has been constituted historically as a governable problem – a genealogy of governing. As Rose (1999:57) states, 'genealogies of government seek to establish the singularity of particular strategies within a field of relations of truth, power and subjectivity by means of a work on symptoms'. This work of diagnosis involves a reconstruction of various historical *problematizations* of governing. These might be defined as the specific situations or historical moments 'in which governing is called into question, in which actors and agents of all sorts must pose the question of how to govern' (Dean, 1999:27). The analysis of problematizations of rule enables the strategies, tactics, programs and authorities seeking to order change to be rendered visible. Most importantly, it allows the historical contingency of governmental strategies to be revealed as events, thus exposing the local, discontinuous, disqualified, illegitimate knowledges that have been hierarchized, ordered or excluded in the name of establishing a 'truthful' knowledge of society (Foucault, 1980:83). Archival material, parliamentary documents, secondary data and semi-structured interviews comprise the forms of evidence used to construct such a genealogy of governing.

'Political' Documents

This book analyses the ways in which agricultural change was governed as a problem from the late-1960s. Numerous documents – those defined as both political and non-political – were used as a means to access both erudite knowledges, and accounts of the accidents, chance events and illegitimate knowledges marginalized or excluded in attempts to unify change as a knowable and governable object. Historical documents represent an important part of a genealogical method in that they provide access to material that may be difficult or impossible to gain via interviews or observation (eg. Dean, 1994). As this project covered a 30 year period, documents were essential in gaining detailed insights into historical events, policies, and debates. An outline of the historical documents utilised is provided in Appendix A.

Semi-Structured Interviews

Changes in the governing of agricultural change are contained not only within material documents, but also in the discursive practices of those authorities who have played a key role in the operation of rule. 'Semi-structured' interviews (Minichiello, Aroni, Timewell and Alexander, 1990; Glesne and Peshkin, 1992) were thus used also as part of a genealogical methodology to provide a means of gaining additional access to those rationalities and technologies of rule already evident in political documents. It should be emphasised that the purpose of interviewing various authorities was not to look for some underlying meaning or truth about the governing of agricultural change, nor to make the statements contained in the material documents more (scientifically) valid through triangulation, as some writers advocate (eg. Denzin, 1978). Rather, the aim was to *crystallize* the statements obtained through the documents, and to access a range of discourses that would enable 'a deepened, complex, thoroughly partial, understanding of the topic' (Richardson, 1994:522). In other words, the interviews were used to gain access to the contemporary rationalities and technologies deployed by authorities – not as a means of finding out how farmers were 'really' governed.

Interviews were conducted in 1997 and 1998 with 16 people who had been involved directly in the Rural Adjustment Scheme within the previous three years. Of these, seven were Commonwealth public servants (Department of Primary Industries and Energy, Rural Industries Research and Development Corporation), four were Queensland State public servants (Queensland Department of Primary Industries, Queensland Treasury, Office of Rural Communities, Queensland Rural Adjustment Authority), and five were representatives of farm organizations (National Farmers' Federation, Queensland Graingrowers, United Graziers Association, Cattlemen's Union). Interviewing and sampling procedures are outlined in Appendix B, while the questions used as the basis of the interviews are listed in Appendix C.

Secondary Data

In addition to official documents, and interviews, the project also utilised selected data from a recent drought study in order to examine some of the ways in which the 'local' knowledge of farmers may not accord with the rationalities and technologies used by authorities to govern them as 'viable' or 'unviable' producers. The study, entitled *Drought in the 1990s: Australian Farm Families' Experiences* (Stehlik, Gray and Lawrence, 1999), was funded by the Rural Industries Research and Development Corporation. It involved comparing the experiences of farm families in two regions affected by drought (1989-1996) – one in Central Queensland (beef producers), and the other in the Riverina region of New South Wales (sheep/wheat producers). Fifty-five adults were interviewed in Central Queensland and 54 in New South Wales. There was a relatively even gender balance in both samples with men and women having been interviewed separately. The research instrument involved both closed and open-ended questions and sought to gain participants' experiences of the drought during 1995/96.

It was the series of interviews with 55 Central Queensland primary producers that formed the basis of the secondary data analysis for this book. Seven questions were selected from the interview that related to the ways in which farmers managed their properties. These questions are outlined and analyzed in Chapter Five. The collected data are consistent with a genealogical methodology in that they provide access to the local knowledge of farmers. Local knowledge is a crucial ingredient in understanding the complexity of farming practice (eg. Kloppenburg, 1991; Long and Villareal, 1994; Portela, 1994; Clark and Murdoch, 1997), and particularly in drawing attention to why farmers adopt, reject, or partially adopt information and technology. Existing data on the knowledge of drought-affected farmers is used in Chapter Five to show the limits of contemporary forms of governing agricultural change. The material is a supplement to political documents and interviews, and seeks specifically to document the limits of governmental strategies that govern through a rationality of managed risk.

OUTLINE OF THE ARGUMENT

Constructing a genealogy of the governing of agricultural change, focusing on specific problematizations of rule, commences in Chapter One where the emergence of a 'low-income farm problem' is discussed. This chapter examines the period from the late-1960s to 1973, and argues that the practices of economists in bringing heterogeneous elements – such as farm incomes, debt levels, theories of agricultural development, and commodity prices – into a unified calculable domain, enabled previous collectivist forms of governing farmers to be called into question. However, tensions emerged throughout this period between the desire to maintain collective forms of support to farmers and the increasing recognition that this was simply adding to the low-income problem. This led to a search for more 'rational' ways of managing the conduct of farmers and an increased political role accorded to economic expertise.

Chapter Two examines the period 1973 to 1985. This period of Australian history was characterised by the belief that assistance to farmers represented a threat to economic management if not governed in a 'rational' manner. The creation of the Industries Assistance Commission (IAC) under the Whitlam Labor government in 1974 marked an important shift in the evaluation of agricultural assistance. Rather than assistance evaluated according to its capacity to maintain the economic welfare of particular industries, such assistance now had to be assessed in terms of its effects on the efficiency of the entire economy. Significant here was the enhanced authority accorded to economists in the preparation of 'professional' submissions, and in transforming the concerns of government and farm organizations into a technical discourse. Nevertheless, as argued in Chapter 2, despite the shift to rational economic management, the conservative Liberal/National Country Party Coalition, in government between 1975 and 1982, sought to shape this form of governing in a social way. Hence, there remained a concern with distributing resources as widely as practicable. This was particularly evident among State authorities where there existed considerable variation in the determination of viability.

Chapter Three examines the period in Australian politics from 1986 to 1989 and the forms of governing that enabled the emergence of farm business capacities as a so-called problem in achieving improved economic growth. It takes as its starting point the constitution of economic crisis that emerged through Treasurer Paul Keating's infamous statement that Australia was in danger of becoming a 'banana republic', and subsequent political moves to internationalise the Australian economy. Of particular interest in this chapter is how previous collectivist forms of governing were constituted as inhibiting outward-looking attitudes on the part of farmers. The linking of an economic crisis to the attitudes and capacities of farmers enabled managerial skills to be constituted as the legitimate sites of restructuring. In this way, the encouragement of improved attitudes via business skills assumed central prominence in defining how farmers should 'properly' respond to change.

Chapter Four shows how a concern with making farmers more internationally competitive was manifested in the focus from the early 1990s with self-reliance and managed risk. These strategies are argued to constitute farm productivity and environmental sustainability as individual business management issues. This means that the capacities of farmers are constituted as a key site of risk, seen as properly addressed only through formal training in business techniques and natural resource management. In fact, as Chapter Four notes, formal training through programs such as Property Management Planning (PMP) and FarmBis, represent specific forms of governing through which farmers are transformed into 'active' managers of risk, and equipped with the attitudes and capacities to manage their properties in a prudent and thus 'viable' manner. Those producers who are unable or unwilling to adopt a more individualistic approach to risk management are targeted as failing to use the available opportunities and constituted as unviable.

Chapter Five extends the concerns of the previous chapter by examining three key sites of contestation to managed risk: exceptional circumstances provisions, local farmer knowledge, and the issue of sustainability. The purpose of this chapter is to draw attention to the specific sites through which advanced liberal rationalities of managed risk encounter limits, and are called into question. Finally, Chapter Six draws together the arguments of the book, evaluates the issues raised, and makes conclusions concerning the regulation of agricultural change. This chapter assesses how various problematizations of governing agricultural change were constructed in Australia between 1967 and 1997. Specifically, five key conclusions are discussed which demonstrate the shifting nature of the relationship between economic expertise and agricultural change as a knowable and governable object of knowledge.

Chapter 1

THE EMERGENCE OF THE 'LOW-INCOME' FARM AS A NATIONAL ECONOMIC PROBLEM

This chapter commences the task of constructing a genealogy of agricultural restructuring locating the emergence in the late 1960s of farm productivity as a distinct national *economic* problem. It focuses particularly on how diverse events, theories and forms of calculation came to constitute a 'low-income farm problem' on which only economic expertise could make authoritative statements. The chapter also reflects on how increased authority of such expertise enabled the questioning of a welfarist rationality of governing that had underpinned agricultural assistance from the 1950s.

The post-War years in Australia were characterised by considerable prosperity. Under the long leadership of Liberal Prime Minister Robert Menzies, from 1949 until 1966, there was an increase in social spending and a belief that state expenditure would create an equitable and just society (Graycar, 1983). At the same time, the Country Party, in Coalition with the Liberals, was gaining prominence under the leadership of John 'Black Jack' McEwen – leader from 1958. The Country Party endorsed strongly 'protectionist' policies that supported small commodity producers, and it favoured government incentives, provided from the early 1950s, to expand agricultural production (Lawrence, 1987). In addition, policies of 'closer settlement' were encouraged by various governments (see Gruen, 1990), in the belief that Australia's vast inland areas needed to be populated to protect its people from the threat of invasion from 'Red Asia'. This was summed up in the slogan 'populate or perish'.

However, by the late 1960s it was becoming clear that the general prosperity of the post-War years was coming to an end. Applied to agriculture, a continuing cost-price squeeze on farmers, the declining contribution of agriculture to the GNP, the impending membership of Britain with the European Economic Community, declining wool prices and diminished access to some overseas markets (Hefford, 1985; Lawrence, 1987; Gruen, 1990) raised doubts concerning previous policies. Following the retirement in January 1966 of Liberal Prime Minister, Robert Menzies, a number of leaders followed in relatively frequent succession including, Harold Holt (1966-1967), John Gorton (1967-1971) and William McMahon (1971-1972). These leaders, while remaining committed to previous policies of subsidization, were faced with the prospect of declining farm incomes, and evidence that policies of closer settlement and agricultural expansion had failed to achieve their objectives.

It was within this political environment that a problem of low-income farms emerged. However, as I argue in this chapter, a tension existed between attempts to take action on the emerging issue of low farm incomes, and objectives of ensuring that farmers had the opportunity to remain on their properties. This tension was manifested in attempts by authorities to provide support to 'marginal' farmers, while pursuing strategies that tended to undermine these objectives. It was evident also in State authorities conducting assessments of viability in ways that suited their own objectives of stabilizing farm incomes. In order to appreciate why the objective of 'restructuring' farming became an important political issue in the late-1960s, it is necessary to first provide a brief historical analysis of how farm policy was shaped and regulated during the immediate post-War period.

The Historical Context in which a National Farm Problem Emerged

The period between 1945 and the late-1960s was one of general stability and prosperity in Australian society and agriculture. The experience of the Great Depression during the 1930s had convinced farm leaders and governments that rural industries needed protection from the free play of market forces (Gruen, 1990). Equally, the Second World War demonstrated that the Commonwealth government could exert some control over the Australian economy. The constitutional powers that the Commonwealth had gained during the War were subsequently used to stabilize agricultural production through controls over prices and supplies. It was thought by authorities that 'a stable agriculture would provide the raw materials for emerging manufacturing industries which would, in turn, provide the basis for population increase' (Lawrence, 1987:186). This objective was further encouraged through policies of 'closer settlement' (see Cockfield, 1995) that involved the subdivision of farming lands believed to be too large for agricultural production. Despite a significant rise in commodity prices due to pent-up demand following the War, stabilization policies ensured that farm investment was restrained by shortages of manufactured inputs and other materials to increase output. As Hefford (1985:81) notes, 'Over the period 1945-50, investment was barely sufficient to make good the disinvestment that had occurred during World War Two'.

However, the Korean War changed the focus of agricultural policy dramatically. The wool boom of 1950-51, brought on largely by the war in Korea, led to a huge increase in exports, but also to a concurrent demand for imports in order to facilitate increased production. This set in motion an inflationary spiral that culminated in a balance-of-payments crisis during 1951. The Commonwealth government believed that this crisis could best be addressed via the introduction of import restrictions, and a program of increased exports through agricultural expansion (Hefford, 1985; Gruen, 1990). The objective of increased output was pursued through welfarist technologies of governing – that sought to collectivize risk – such as tax concessions, subsidised prices, low interest credit, and bounties. Assistance to agriculture rose throughout the 1950s and 1960s with output rising by more than 35 per cent (Gruen, 1990). Such forms of assistance were mirrored in general policy-making where a rationality of governing based on government-directed economic growth, and guided by the theories of British economist John Maynard Keynes, was predominant (Smyth, 1994).

The policy objective of agricultural expansion, however, began to be called into question from the mid-1960s. Four reasons are typically provided concerning why the performance of some farms was at this time seen to be in need of serious reform. First, the claimed 'normal' processes of economic development resulted in a decline in the importance of agriculture and contributed to a relative fall in farmers' terms of trade (Standen and Musgrave, 1969; Harris, 1970). Second, the increasing economic importance of mineral development and the tertiary sector during the 1960s caused agriculture's share of GDP to gradually decline (Kingma and Samuel, 1977; Lawrence, 1987; Gruen, 1990). Third, as output increased relative to underlying demand conditions, greater government support was needed to equalize returns, and the rationale for such assistance came under increased scrutiny (Industry Commission, 1996:18). Fourth, and finally, as farmers increased their output, the prices which they received for their commodities tended to fall relative to the cost of inputs: a 'cost-price squeeze' (see Standen and Musgrave, 1969; Lawrence, 1987; Gruen, 1990). While some farmers could afford to invest in capital to become more efficient, and thereby offset declining terms of trade, many small commodity producers found themselves fighting an uphill battle to increase output. These explanations for a crisis in expansionist policies are all compelling. However, they do not account for *how* these events and reflections on governing enabled previous forms of intervention to be problematized as a unified low-income farm problem.

While some agricultural economists had long warned of problems arising from closer settlement and policies of expansion (see Edwards and Watson, 1978), it was not until a 1967 keynote address by the President of the Australian Agricultural Economics Society that welfarist forms of agricultural regulation began to be seriously called into question (see McKay, 1967). In fact, as I argue below, agricultural economists were instrumental in constituting low-income farms as a problem with welfarist strategies of governing. This was more than a simple exercise of describing in a neutral manner pre-existing problems. Rather, it required 'Australian agriculture' to be represented in a specific way. For farmers' viability to be rendered a problem, an agricultural sector with particular characteristics, limitations and interests first needed to be made visible and brought into a calculable domain (see Murdoch and Ward, 1997 for a similar argument in the context of UK agriculture). Through the use of theories, statistics and types of calculation, it was possible for knowledge concerning a low-income farm problem to be accumulated and assembled as a distinct 'issue'. The constitution of the low-income farm problem through the agricultural economics literature will now be outlined.

THE CONSTITUTION OF THE 'LOW-INCOME FARM PROBLEM'

From the early 1960s a number of agricultural economists had noted the existence of farms earning what were considered to be low-incomes (eg. Lewis, 1961). However, the evidence collected suggested that these farms were concentrated in dairying (eg. Bird and Marriott, 1966), and a low-income problem was therefore seen as confined to that industry. It was not until 1967 that an important shift in rationalities of governing occurred that constituted low-income farms as part of broader economic trends, having the potential to affect all industries. This problem was constituted primarily through the work of agricultural economists in the Federal and State public service, and at the rural University of New

England in Armidale (New South Wales), who were attempting to understand this 'new' issue and diagnose solutions.

Genealogically, the works presented below typically proceed through two stages of staking their claim to truth. First, they attempt to link farm incomes, as a truthful representation of farm practices, to a broader (economic) problem of agricultural viability. Second, they diagnose how this problem might best be governed. Both of these stages attempt to establish a 'truthful' account concerning such things as what a viable farm is, how a set of processes called an economy operates, and what aspects of farm conduct require, and are most amenable to intervention.

'The Small Farm Problem in Australia' (McKay, 1967)

A shift away from conceptualizing the low-income problem as confined to dairying was signalled in a paper delivered in February 1967 by the Director of the Bureau of Agricultural Economics, D.H. McKay, to the Agricultural Economics Society. The Bureau of Agricultural Economics (BAE) was an agency of the Commonwealth government dedicated largely to the gathering of information concerning trends in commodity prices, and providing agricultural policy advice to Federal agencies and Ministers. Therefore, the BAE was recognized formally as an agent of economic expertise, and McKay's paper was accepted as an authoritative form of knowledge. Moreover, as McKay was a senior public servant, his views may be regarded as acceptable to the Liberal/Country Party Coalition government of the day.

In his paper, McKay sought to focus primarily on disparities between incomes in agriculture, and other industries. Through the use of statistics and graphs, he argued that while average farm incomes were in line with non-farm incomes, there were substantial differences between industries. High incomes in the wheat, beef and wheat-sheep industries tended to skew the overall figures. In industries such as dairying and horticulture, incomes were far below the 'average'. For McKay, this indicated the presence of not only a group of producers who faced immediate problems, but also a significant number who 'may well become a problem' (McKay, 1967:130). Using the arbitrary figure of $2,000 per annum as the dividing line for a viable farm income, McKay (1967:124) estimated that 80,000 or one-third of Australia's 250,000 farms fitted into the 'low-income' category. Of these, 34,000 were in the dairy industry. By way of comparison, a survey in 1966 headed by R.F. Henderson (who was later commissioned by the Commonwealth government to report on poverty in Australia) of 4,000 people living in Melbourne, argued that an annual income of $1,716 constituted an urban 'poverty line' (Henderson, Harcourt and Harper, 1970). In a similar manner, McKay's estimate might be considered a poverty line for farmers.

Despite recognising the limitations of his conclusions McKay, nevertheless, sought to identify the causal factors contributing to this low-income problem. These included first, the physical constraints of size that inhibited the level of farm output; second, the lack of managerial ability on the part of farmers; and third, the inefficiencies introduced by the inability to purchase new capital to increase farm output. McKay's mentioning of farm management is notable given that it was not seen as a significant part of governing change until the late 1980s. This indicates that advanced liberal rationalities of farmer self-management (see Chapters Four and Five) were circulating within the agricultural bureaucracy well before they took a more material form in government policy.

In identifying the 'causal factors' of change, McKay problematized the difficulties facing some farmers as a low-income farm problem and, in doing so, linked a number of heterogeneous factors into a unified technical form. This was defined in a narrow way and tended to ignore factors such as perquisites and way-of-life considerations (Edwards and Watson, 1978). From the perspective of an analytics of government, however, the rendering of farm incomes into a comparable and quantifiable statistical form enabled a stable representation of the low income problem to emerge which could then be made amenable to 'evaluation, calculation and intervention' (see Miller and Rose, 1990:7; see also Murdoch and Ward, 1997).

The emergence of agricultural change as a low-income problem was made possible through a calling into question of welfarist rationalities of governing. This was evident in McKay's questioning of subsidization as a means of governing the productivity of farmers. McKay (1967) perceptively noted that it could not be guaranteed that the political will would always exist to make increasing transfer payments to farmers. Moreover, subsidization was argued to benefit high-income producers and thus exacerbate the low-income problem. In concluding his argument, McKay stated:

> I find myself driven to the conclusion that we in Australia must look more closely than we have in the past at the structural adjustments which may be required in the agricultural sector to ameliorate the low income problem (McKay, 1967:132).

While this statement transformed the previous disparate speculations into a unified problem worthy of political attention, McKay's final point questioned decisively subsidization and welfarist forms of intervention:

> I believe we will have to consider very carefully whether the structural adjustments required should be left to the economic forces working on rural industries or whether the adjustments will require more positive steps to reduce the number of low income farms (McKay, 1967:132).

As noted in the previous introductory chapter, governance approached genealogically is a problematizing activity. In the above statements, McKay was not suggesting that farmers facing difficulties should or should not be governed. Rather, he was posing a classic problem of liberalism: *how much* should one govern? This question formed the basis for the deliberations of the following group of agricultural economists examined.

'The Agricultural Adjustment Problem with Specific Reference to Australia' (Standen and Musgrave, 1969)

By 1968 economists and politicians alike had begun to accept the idea that many identified problems, such as the 'cost-price squeeze', were not transitory in nature and required government intervention. However, how best to intervene was still a matter of some debate. A National Workshop on Agricultural Adjustment Problems conducted at the University of New England in Armidale, New South Wales, in February 1968 was an attempt to better understand the low-income farm problem – identified by McKay – and thus provide directions on how it might best be studied and addressed.

The workshop provided an opportunity to share ideas, and was dominated by agricultural economists from universities and the public service. Perhaps not surprisingly, its broad conclusions sought to promote a more 'rational' efficiency-based approach to the governing of agricultural change. This type of rationality was summed up by Lewis who stated:

> ...assault(s) on economic rationality leave the door open to pressure group politics which can make Australia much poorer and, indeed, may be doing so... Our institutions must be adapted to introduce economic reality to public policy decisions unless we are willing to accept a seriously sub-optimal rate of economic growth. (Lewis, 1969:231)

As members of a relatively young sub-profession, agricultural economists were striving to stake their claim to scientific legitimacy, as were economists more generally (see Smyth, 1994). The statement by Lewis attempted to authorize the expertise of agricultural economists as crucial in averting 'pressure group politics', and promoting a more economically rational means of governing. It is not my intention to examine all of the papers presented at the 'Problems of Change' Workshop. Only one will be analyzed in some detail, that entitled 'The Agricultural Adjustment Problem with Particular Reference to Australia' by B.J. Standen and W.F. Musgrave. This paper has broad political significance as it was drawn upon subsequently by a number of economists writing on the governing of agricultural change (eg. Harris et al, 1974; Kingma and Samuel, 1977).

Essentially, Standen and Musgrave (1969) sought to use the influential arguments of United States agricultural economists T.W. Schultz and E.O. Heady to establish whether a farm problem existed in Australia and, if so, what types of government intervention might be considered. It is crucial to note that these views were of great significance in shaping how agricultural economists reflected upon the farm problem in Australia. Briefly, Schultz (1949) argued that in considering solutions for the economic problems from growth in agriculture, it was necessary for governments to focus on the longer run effects of price instability on farm incomes, capital and land. It might be expected from the perspective of neo-classical economic theory that when farm earnings dropped to lower than incomes in urban areas, farmers would make a decision to leave their property and take higher financial rewards in towns and cities. However, in practice this was not occurring. For Schultz, this gave rise to a farm problem in which resources were not being allocated to their most efficient use, and thus a *farm problem* in which a role for government was necessary.

Heady (1966) developed Schultz's ideas during the 1960s. His thesis, as applied by Standen and Musgrave, can be summarized as follows:

- While consumer demand for food is relatively inelastic, changing resource prices and advancing technology make supply elastic.
- Increased supplies of capital due to economic growth make labour more expensive resulting in pressure for the substitution of capital for labour and a 'cost-price squeeze'.
- This results in the development of a farm structure where fewer farmers work on larger farms with more capital intensive technologies.
- The 'cost-price squeeze' places financial pressure on smaller, less capital intensive farmers creating a process of structural adjustment.

- This does not necessarily create a low income problem unless the movement of labour and capital out of agriculture is insufficient to allow income parity with resources in other sectors.
- Government policies should therefore be aimed at facilitating the movement of labour and other resources out of agriculture so that they continue to receive satisfactory returns and so that farmers are compensated for capital and assets which are unprofitable to turn to other uses. (derived from Standen and Musgrave after Heady, 1966)

Overall, Standen and Musgrave (1969) endorsed Heady's thesis in terms of its recognition of farm incomes as a long-term resource efficiency problem, and castigated those policies 'directed at shielding agriculture as though it were troubled by some temporary disadvantage' (18). They noted, however, that the pressures for adjustment in Australia had not been as severe as in other countries such as the United States. This was attributed to:

1) an export orientation which resulted in elastic demand for commodities;
2) the historical land base where large amounts of land were occupied for grazing purposes, and
3) the low prices at which Australian farmers were willing to produce export commodities (Standen and Musgrave, 1969:25).

This was not taken to mean an absence of an adjustment problem in Australia. In a similar manner to McKay, Standen and Musgrave observed a concentration of low-income farms in dairying, fruitgrowing and high rainfall wool production. Many enterprizes in these industries were argued to be too small to equalize returns in the face of falling prices. These industries were noted to contain many of the characteristics outlined in Heady's thesis and were held to be representative of Australia's farm problem. Standen and Musgrave concluded their paper by problematizing government intervention in agricultural change and asking whether the low incomes in some parts of Australian agriculture might be indicative of a temporary period of reduced returns, or whether broader 'structural' changes across agricultural industries were required.

While not detailing an explicit role for government, the authors argued that more serious attention needed to be given to increasing quantities of capital, larger farms and greater opportunities for obtaining credit, so that 'efficient' resource use could be promoted. From the perspective of an analytics of government, the implications for governing were made clear: claimed natural economic processes were being disrupted through inefficient resource use; and since efficient resource use was essential to the economic security of the population, the task of good government was to intervene and promote the optimal distribution of resources. This view of agricultural regulation, drawing upon the Schultz-Heady thesis, continues to play a crucial role in the contemporary management of agricultural change (see Malcolm et al., 2000). It was not until 1970 that a more explicit role for political authorities in addressing the low-income farm problem began to be articulated.

'Farm Adjustment and the Role of Government' (Harris, 1970)

On the basis of its subsequent use in texts on agricultural policy it seems that Harris' work on the role of government – published in the *Australian Journal of Agricultural Economics* – had the most significant impact on the governing of agricultural change throughout the 1970s. As Harris was at the time of the article McKay's successor as the Director of the Commonwealth Bureau of Agricultural Economics, it can be assumed that his views on both the low-income problem and possible options for intervention were largely acceptable to the Federal government (Hefford, 1985:270-271).

Similar to the previous agricultural economists examined, Harris sought to establish whether the existence of low-income farms represented a problem of national economic significance. He commences his analysis by explaining why change is needed in the farm sectors of industrialized nations. Like Standen and Musgrave, Harris used the Heady thesis to explain the emergence of low-income farms in Australia. As noted above, the Heady thesis argued that low-income farms emerged as a consequence of inadequate adjustment to changed economic conditions. While Harris noted that the incomes of Australian farmers generally remained comparable with the non-farm sector in the post-War period, this situation had deteriorated from the late 1960s, due to farmers not transferring to alternative occupations when their incomes reached low levels. Harris concluded that a farm income problem indeed existed (Harris, 1970:130).

However, McKay's (1967) arbitrary figure of $2,000 was rejected by Harris as an accurate basis for judging whether a farmer was earning an adequate income. For Harris, not all farms earning around $2,000 were problem farms or were facing a situation of poverty. Equally, comparisons between the welfare of farm and non-farm families were seen as limited in the conclusions they provided. Nevertheless, as Harris stated:

> The important thing is that a substantial proportion of Australian farmers appear to be earning incomes too low to give them much chance of financing the further investment necessary to enable them to maintain income parity with the rest of the community in the future (Harris, 1970:111).

The point here was that farmers were remaining on their land well after their incomes had gone below what was considered then the 'poverty line' for an urban family (see Henderson et al., 1970). There are a number of economic and cultural reasons why so-called marginal farmers choose to remain on their properties, some of which are outlined in Chapter Five (see also Lawrence, 1987 for a sociological analysis). Given that farmers deemed to be below the poverty line were choosing to stay on their land, Harris (1970:114) argued that 'on the grounds of economic efficiency...a case exists for government policies to deal with the low income question because the economy could benefit by using the resources elsewhere'.

It is interesting to note that despite its references to improving economic efficiency, the rationality of governing represented in Harris' article was decidedly welfarist, in that it sought to achieve economic objectives through the promotion of social responsibility. That is to say, economic efficiency (and thus national growth) was seen as possible through first ensuring that the welfare of low-income farmers was comparable to non-farm populations. Parity of income was thus to be promoted through the bureaucratic capacities of the state. By the same token, however, it would be naive to assume that the specific forms or technologies of intervention suggested by Harris were the same as subsidization. While rationalities of governing remained animated by a will to govern through 'the social', the technologies to achieve this were subject to increasing question. This is borne out in a statement by Harris in which he notes that the purpose of assistance should be to reduce resistance to change from those most likely to lose out, and 'speed the rate at which change is accepted' (Harris, 1970:114). Such a means of acting on farm practices is different from that underpinning intervention in the post-War period, in which the purpose of assistance was to guard against the risks of change by collectivizing the risk of *all* farmers.

Consistent with this shift, Harris concluded his article by suggesting policy objectives that might potentially underpin government involvement in addressing the low-income problem. These included strategies to facilitate voluntary amalgamation of small farms, facilitate upgrading of low income farms where feasible, facilitate departure of low income farmers and avoid development of new low income farms (Harris, 1970:116-117). At the time that Harris was writing, a Commonwealth Liberal/Country Party Coalition government, under Prime Minister John Gorton, had already initiated measures to address the low-income situation in the dairying industry through the Marginal Dairy Farms Reconstruction Scheme, discussed in more detail later in this chapter. Similar to his predecessor McKay, Harris argued that problems of low farm incomes were more widespread than just one sector, and that there was no reason 'why a low income wheat, sheep, dried fruit or even beef producer should not have the same opportunities available to him [sic] as a dairy farmer' (1970:118). Harris here constituted low-income farms as a social problem that needed to be assisted by government if national growth and economic security were to be maintained. Moreover, he based his arguments on a welfarist rationality of governing in which all agricultural industries were argued to be entitled to equity of government assistance.

The three papers discussed above make it clear that agricultural economists played a crucial role in the constitution of agricultural change as a problem of low-income farms. Particular assumptions underpinned how this problem was constituted. First, that a 'cost-price squeeze' in agriculture was a direct result of successful economic growth in industrialised nations. Second, both a welfare and efficiency problem arose if farmers did not adjust to these pressures for change. Third, it was the duty of governments to intervene in order to maintain the economic security of both farmers and society. The posing of questions concerning the limitations of welfarist forms of governing contributed to an increased concern with farm 'reconstruction' from the early 1970s. Nevertheless, as argued in more detail below, the rationalities on which reconstruction was based still had as their objective the fostering of welfarist forms of governing, even if the technologies used to achieve them contradicted such an aim.

Rural Reconstruction and the Regulation of Low-Income Farms

Low-income farms had been identified in the dairying industry since the early 1960s, although this was not then defined as a problem relevant to other industries. Due to key statements by McKay (1967) and others the Commonwealth government by the late 1960s had largely accepted that there was a low-income problem requiring intervention extending beyond the dairy industry (see for example Harris, 1970:116). This involved some strategic political re-alignments. As part of a conservative Coalition, the Liberal Party had sought to leave rural affairs to its governing partner, the Country Party, on whose support it depended. However, in early 1970 the Liberal Party, under the leadership of John Gorton, issued a statement of its own policy guidelines for rural industries that included endorsement of assistance measures to facilitate the exit of low income farmers (Hefford, 1985:70). The reasons justifying this move were provided by Anthony Street, Liberal Party Member for Corangamite (Victoria), who argued during the second reading[1] of the 1969 Appropriation Bill (Budget debate) that:

> It is becoming clear that all Australian industry will have to be prepared for structural change and rationalization. Australia will have to keep up with the world since the world will not slow down for us. The Government will have to realise that it will be much cheaper in the long run to encourage restructuring measures financially now, rather than drift on until the position becomes acute. (Commonwealth of Australia, 1969:475)

Such a position was at odds with the objectives of the Country Party that sought to promote itself as the farmer's party, and maintain measures designed to protect the farm sector – particularly under the leadership of the notorious John McEwen, who was Country Party leader until February 1971.

McEwen's opposition to the Liberal Party's endorsement of farm sector reform contributed in part to Gorton losing the leadership to William McMahon in March 1971. Following the presumed drowning of Prime Minister Harold Holt in 1967[2], McEwen had declined to accept McMahon as leader of the Liberal Party – hence the election of Gorton. However, following a seven per cent swing against the Liberal/Country Party Coalition government in the 1969 elections, McEwen withdrew his veto over McMahon so that McMahon was able to become Prime Minister (Craig, 1991:248). McEwen subsequently endorsed the introduction of reconstruction legislation concerning the dairy industry in May 1970. However, it would take a new Country Party leader (Doug Anthony) to approve the extension of reconstruction assistance to the sheep-wheat industries in 1971.

Marginal Dairy Farms Reconstruction Scheme (1970)

Plans had been mooted for reconstruction of the dairy industry since 1966 when the Dairy Industry Council had recommended that the Federal government provide funds through the States for farms that were uneconomic in size. However, legislation was not passed until mid-1970 by which time most States had implemented for their dairy farmers various forms of assistance. Up until this time, butter and cheese products were subsidized by the Federal government. Falling prices and rising costs during the 1960s had resulted in a fall in farm incomes and a questioning of subsidies as the basis for maintaining farm incomes.

As noted earlier, research by agricultural economists had identified the existence of farms that were not earning what were considered to be sufficient returns (Bird and Marriott, 1966;

[1] The term 'second reading' refers to where a Bill is formally debated in Parliament. All statements by Members of Parliament are in this book drawn from second reading speeches.

McCarthy, Hodgson and Briggs, 1966; Bird, 1969). Moreover, this was also deemed to be a regional problem as most low income farms were concentrated on the northern coast of New South Wales and in south-east Queensland (eg. Cassidy, McCarthy and Hamilton, 1969). The Marginal Dairy Farms Reconstruction Scheme was therefore formulated as a response to these problems. It involved an advance of $25 million to the States (primarily Queensland and New South Wales) to be used over four years. Half of this sum was to be used for administration purposes with the balance to be repaid to the Commonwealth with interest of six per cent per annum.

However, in responding to the difficulties of some dairy farmers, this Scheme sought also to constitute the low income problem and make it amenable to intervention by seeking to differentiate between viable and unviable farmers, and to establish categories through which these farmers could be assisted. Rather than increase subsidies for farmers, as was the case previously, this Scheme had two key objectives:

- to enable low income dairy farmers who voluntarily wish to do so, to leave the industry and to receive a fair price for their land and improvements; and,
- after the writing-off of redundant assets, to make the land and useful improvements available to other farmers so as to build up their properties to a viable family farm level and, where possible, diversifying the pattern of land use (Commonwealth of Australia, 1970a:2187).

The main aim of these objectives was to ensure that those farmers wishing to leave the industry were not disadvantaged and, that those farmers taking over the land would improve productivity and thus their income. However, the opposition Labor Member for Wilmot (Tasmania), Gilbert Duthie, noted that there was nothing to stop a dairy farmer purchasing a neighboring low income farm and then proceeding to increase production (Commonwealth of Australia, 1970b:2924). This was argued to defeat one of the purposes of the Bill, which was diversification in the interests of reducing national butter surpluses. Nevertheless, 'getting big' via the amalgamation of farms, or 'getting out', was assumed by the Federal government to represent the best means of improving efficiency in the dairy industry.

Unlike later schemes, the Commonwealth did not provide a Schedule as to how the States should administer the funding or the specific policy instruments that should be used. Only broad guidelines were provided. It was therefore up to the individual State authorities to decide how farmers might be either assisted off their land, or be given funding to build up their farms. As long as the States followed the objectives of facilitating farm build-up, and diversification and assistance for those farmers exiting the industry, it was up to them as to how this process took place. The *Marginal Dairy Farms Agreement Act* (1970) thus allowed the making of agreements, subject to certain conditions, between the Commonwealth and States. For example, in all agreements a 'marginal' dairy farm was to be defined as one running a minimum of twenty lactating cows with over half the income coming from the production of milk or cream, but one with an 'overall production insufficient to provide a reasonable level of income' (Commonwealth of Australia, 1970a:2188). This ensured that hobby farmers were not assisted.

---

[2] Harold Holt disappeared, presumed drowned, after a swimming trip off the coast near Portsea in southern Victoria in December 1967.

More significantly, it sought to make knowable the characteristics of, and thus differentiate between, normal and marginal farms. In doing this, research could then be conducted on 'unviable' cases, with statistics collected and collated into a material form, the characteristics of 'normal' farms represented and confirmed as truthful, and marginal farms made amenable to government regulation. The Marginal Dairy Farms Reconstruction Program remained in place until 1974 when a Dairy Adjustment Scheme was implemented under the Whitlam Labor government. This short-term scheme was terminated in 1975 by the same government, with adjustment assistance for dairying thereafter administered through the Rural Reconstruction Scheme (outlined in a later section of this chapter).

One of the main assumptions – and in retrospect one of its limitations from the perspective of authorities – of the Marginal Dairy Farm Bill was that dairy farmers who were defined as marginal would realize that their income situation was unsustainable and voluntarily seek reconstruction assistance. However, as the Labor Member for Dawson (Queensland), Rex Patterson, stated it was not this simple:

> I feel that the primary producers, particularly the dairy industry farmers and their families in marginal areas, are so entrenched in dairy production, which is their way of life, that even with this voluntary system it will be very difficult to entice many of them from their farms. We should not criticize them for wanting to remain on their properties because this is their way of life. They want to live this way. Although possibly they may not receive as much as the basic wage, many of them are happier than other people who work in the city. (Commonwealth of Australia, 1970b:2910)

This statement raised doubts concerning the capacity of government programs to effectively encourage those farmers defined as unviable to exit the industry. Thus, despite the Bill having agreement by all parties, and being assented to on the 23rd of June 1970, some members in the Opposition expressed reservations concerning how the legislation would be governable in practice if it relied on the choices of farmers to leave. This is a significant issue that will be revisited later in this chapter. At this point, however, it is necessary to outline how reconstruction became constructed as a broader problem that extended beyond the dairy industry to other agricultural activities.

Australian Rural Economy – Parliamentary Debate (August 1970)

As noted above, a number of disparate events were believed by many politicians and agricultural economists to be indicative of a need for reconstruction within certain agricultural industries. However, apart from the dairy industry, these ideas had not been applied practically to other industries. This changed in August 1970 when the Country Party Minister for Primary Industry, Doug Anthony, commissioned the Bureau of Agricultural Economics (BAE) to investigate, and report on, the need for immediate and longer term adjustment with specific reference to the wool industry (discussed below).

At this time, wool was one of the largest industries in Australia and comprised almost half of the total commercial farm holdings in Australia. Up until the late 1960s, the BAE had undertaken largely cost-of-production surveys that were used as the 'raw material' by the Country Party and many farm organizations to support their arguments for price stabilization schemes (Connors, 1996:68). While Anthony was known for his dislike of economists, he was by 1967 beginning to question the tactics of his own Party in using the BAE figures to support the argument for 'protection all round'. On this basis, he gained Cabinet approval in 1967 to extend the largely fact-finding role of the BAE. The BAE's role was enhanced further when Anthony replaced John McEwen as leader of the Country Party in 1971. In fact, as Martin (1990:158) notes, during the early 1970s the BAE changed its role considerably from fact-finding responsibilities to policy analysis and forecasting roles.

The commissioning of a report on farm viability in the wool industry coincided with the calling of an urgent debate on the 19th of August 1970 by the Labor Member for Dawson, Rex Patterson, concerning the state of the Australian rural economy. Dr. Patterson argued that domestic policies of price support in manufacturing, and the poor state of world markets, were having an adverse impact on farm viability. He noted that this was particularly evident in the wool industry where rising costs and falling prices had contributed to a significant drop in farm incomes. 'Reconstruction' was in the subsequent debate discussed as a strategy for addressing this problem and in helping low-income farmers who wished to stay in farming survive.

Significantly, Members from both sides of Parliament agreed on reconstruction as a necessary strategy to improve the income situation of farmers, although there was less of a consensus on what specific form that this might take. As Dr Patterson stated, 'Reconstruction of the wool industry is urgently required. Handouts simply delay the inevitable' (Commonwealth of Australia, 1970c:155). This statement represented a direct challenge to rationalities that sought to govern farmers through welfarist forms of regulation such as subsidization. Reconstruction of agricultural industries via forms of assistance that facilitated the exit of some farmers while improving the profitability of others was seen as a key priority at the Federal level of government – a view now shared by all parties. The BAE report was important in transforming these priorities into a technical discourse and material form through which intervention could be pursued.

Debt Reconstruction and Farm Adjustment (BAE, February 1971)

The BAE Report on debt reconstruction and farm adjustment (BAE, 1971) was released in February 1971. Its arguments were very similar in nature to those contained in Harris' (1970) article. As Harris remained in the directorship of the BAE this was perhaps unsurprising. However, the BAE report sought to extend these ideas and propose possible forms of assistance based upon a rationality of reconstruction. The BAE reinforced the earlier work conducted in the previous decade by agricultural economists in both universities and the public service.

The BAE report argued that adjustment to market conditions was occurring at an unsatisfactory rate, and that high levels of debt and farmers with low-incomes were indicative of longer-term economic problems in Australian agriculture. This was applied primarily to the wool industry, although the Bureau noted that problems of economic viability were applicable to other industries. As the Report (1971:4) noted:

> Debt problems of the magnitude currently being experienced are a sign of longer-term economic weaknesses in parts of the industry structure. Any debt reconstruction measures, to be effective and to lessen the likelihood of similar financial problems arising

> to the same extent in the future, must be associated directly with measures designed to facilitate structural change in the industry.

Rather than being conceptualized as a temporary phenomenon, corrected through collectivist forms of governing – as was the case in the past – the debt situation in the wool industry was here linked to broader economic problems argued to require long-term reform – focusing primarily on the capacities of low income farmers.

In order to establish that a problem existed requiring changes to industry structure, such statistics as farm income and total rural indebtedness by farming zone and debt per sheep equivalent were collated and presented in a tabular form. From these statistics, the debt position of the wool industry was compared with other industries and evaluated. While the BAE report argued that the tabulated figures did not indicate a serious situation in themselves, they were 'consistent with a situation affording concern in significant parts of the industry' (BAE, 1971:6). This was despite the BAE's admission that the evidence regarding debt was 'inevitably limited' and that the application of any single or precise criterion to measure the extent of the debt situation was difficult. Nevertheless, tentative assessments were made regarding the debt situation in the wool industry with producers grouped into three categories:

- Those whose debts represent a large proportion of the total equity value of their property.
- Those whose equity remains reasonable in relation to their total debt but who have difficulty in servicing their debt.
- Those with small debts, but whose incomes are low or may become low in time because they have little capacity to adjust further in the face of continuing income pressures. (derived from BAE, 1971:13-14).

Producers in these categories were constituted as a problem to the future prosperity of the industry, although debt was seen as only one determinant in defining a farms viability. This assessment was extended to other industries with the BAE suggesting:

> In the absence of a significant fall in the number of farm operators, average net farm income per property will decline. At the present time there are too many attempting to earn an income from agriculture, given the market situation, both in the sheep industry and in most other lines of agricultural production. (Bureau of Agricultural Economics, 1971:15)

The BAE argued here that a decline in the number of unviable farmers was a necessary step to alleviate the total debt situation and increase land areas, and thus incomes in both the wool, and other agricultural industries. This was based on the assumption that 'decreasing unit costs may be expected as enterprize size increases' (BAE, 1971:20).

Debt problems in the wool industry were claimed to compromise seriously the efficiency of the sector as a whole, and were used by the BAE to support a case for government action of a debt restructuring and farm adjustment nature. On the one hand, debt reconstruction assistance was regarded as a form of 'rearguard action' (1971:19) to ensure that producers whose enterprises were temporarily having problems, but were 'basically sound' (1971:19) in the long term, had access to long-term loans. On the other hand, farm adjustment assistance was argued to be best targeted at producers with a 'low prospective income earning capacity'

(1971:19) to ease the enlargement of property size and assist with farm amalgamation. Both forms of assistance were deemed necessary to address the efficiency *and* welfare aspects of the debt problem. A role for governments was thereby accepted as a means of distributing industry resources towards their most economically effective use, and ensuring that low income farmers who wished to leave their properties were adequately compensated 'at a fair market price' (BAE, 1971:21). The BAE believed that this would maintain the stability of agricultural industries at the same time as providing low-income farmers with an alternative to further income deprivation.

Evident in the BAE's comments was an assumption that most producers, given sufficient government assistance, would have the capacities to overcome their debt and thus achieve viability in the longer-term. Those farmers defined as unviable were deemed to be candidates for poverty and were thus argued to require assistance to help them find work in another industry. This type of rationality for governing farmers had as much to do with poverty alleviation as it did with increasing productivity. The BAE's key objective was thus the stabilization of farm incomes. It was believed that through reconstruction, productivity could be improved, leading to increases in farm incomes and improvements in the economic welfare of farmers – at least for those remaining in the industry.

In order to facilitate the structural change argued to be required in the wool industry, and other agricultural industries, the BAE suggested the need for a set of practical policy measures. However, an extension of arrangements under the Marginal Dairy Farms Reconstruction Scheme to the wool industry was argued as simplistic given that the problems in wool were more complex and extensive than those in dairying. These included debt reconstruction through the provision of a long term credit facility for those farmers considered to be viable, facilities to assist amalgamation of unviable properties to increase their productivity, and assistance to those farmers who wished, or who were forced by circumstances, to leave the industry. Overall, the BAE argued that the implementation of these objectives required institutional change. While the Marginal Dairy Farms program provided one such example of how this could be achieved, 'structural' problems in other industries were argued to require more complex policy responses. The recommendations of the BAE contributed to the formulation by the Commonwealth Government of a rural reconstruction scheme in April 1971. However, this debate was preceded by a discussion concerning the inaction by government in addressing a claimed financial crisis in Australian agriculture.

Rural Reconstruction – Parliamentary Debate (April 1971)

At the same time as strategies were being considered to provide assistance to the wool industry, evidence was accumulating that pointed to the existence of crises in other industries. It was the existence of these claimed crises that on the 27th of April 1970 formed the basis of the Labor Opposition spokesperson for Primary Industry, Rex Patterson's, call for reconstruction to be given priority on the national political agenda. Patterson argued that rising production and marketing costs necessitated the immediate implementation of reconstruction programs 'aimed at progressively achieving levels of farm production in line with realistic market demand' (Commonwealth of Australia, 1971b:2056). The strategy of addressing the problem of low-income farms through the redistribution of resources to those producers deemed to use them in the most efficient manner was by this time not in dispute among the major political parties at the Federal level of government. Reconstruction was seen as the most rational means of dealing with the claimed structural crisis of falling incomes. The problem from the Labor Opposition's point of view was that legislation had not been implemented fast enough, and that the Government had not stated the specific criteria through which either reconstruction should occur or how many farmers would have to leave.

While the exact number of farmers deemed as needing to leave farming was never stated publicly, the Country Party Minister for Primary Industry, Ian Sinclair, provided a categorization of the producers – based on the BAE's model – that needed to be assisted. This categorization sought to render visible an 'adjustment' problem. It comprised:

- Producers unlikely to be profitable and therefore in need of assistance to help them sell their land and move off their properties.
- Producers who, given a better consolidation of their debts and given a longer term on borrowings, might reasonably be expected to get back in a position of profitability.
- Producers who are not yet in difficulty but who, given the short term period over which their funds are borrowed, may require a re-phasing of their debts over a longer term to ensure they remain in a situation of profitability. (derived from Commonwealth of Australia, 1971b:2059)

In categorizing farmers according to income, profitability and debt levels, the Minister was effectively representing farm viability in a way that made it amenable to intervention. That is to say, in deploying 'income', 'profitability' and 'debt' as claimed accurate representations of farm practice, particular regularities and differences between producers could be rendered knowable and the technical knowledges of experts, such as the BAE, brought to bear on this information in the interests of formulating strategies for governing the conduct of farmers. Such a categorization of farmers was crucial in transforming 'reconstruction' into a material form via the Rural Reconstruction Scheme (RRS).

Rural Reconstruction Scheme (April 1971)

The parliamentary debate on the Rural Reconstruction Scheme commenced just two days after the issue of rural reconstruction had been declared a matter of urgency. One of the main reasons why the legislation – the *States Grants (Rural Reconstruction) Bill* (1971) – was so long in reaching the second reading stage was that the Commonwealth was negotiating with the various State governments to seek their agreement on the administration of the proposed scheme. Once the States had agreed, the Commonwealth could then seek approval to administer the funding of $100 million over a period of four years. The second reading of this Bill was thus a debate seeking parliamentary approval to appropriate Commonwealth funds to provide to the states for rural reconstruction purposes. Specific details of the scheme were contained in a schedule to the agreement at the end of the Bill.

The *States Grants (Rural Reconstruction) Bill* (Commonwealth of Australia, 1971b) was formulated largely as a response to the economic downturn – which was rendered a knowable problem by the BAE – in Australian agriculture, but particularly in the sheep and sheep-wheat industries. Using BAE figures, the price of wool, debt levels and loss of farm income were all deployed as indicators of a crisis that could not be addressed through previously dominant price support policies. As the Country Party Minister for Primary Industry, Ian Sinclair, noted:

> The burden on the budget of a price support scheme of the magnitude required would be enormous. Apart from this it is essential for the health of the industry that it should *not* be fully insulated from the effects of changes in demand as reflected in price (Commonwealth of Australia, 1971c:2333, italics V.H.).

Previous 'protectionist' policies were thus discounted as unrealistic, with budget and market realities argued to require forms of intervention, that exposed farmers to changes in market demand.

This was a radical shift given the Country Party's dogmatic support during the 1960s for assistance that protected farmers from the effects of changes in market prices. The departure of John McEwen in conjunction with the increasing acceptance by Country Party members of the BAE's advice on low-income farms enabled a strategy of reconstruction to be asserted as the most 'rational' means of governing change. This was to be achieved through three forms of assistance outlined below. These forms of assistance accorded with the three categories of 'problem farmers' outlined by the Minister for Primary Industry in the parliamentary debate on rural reconstruction (see Commonwealth of Australia, 1971b), and the BAE.

The forms of assistance included:

- Debt reconstruction – To assist a farmer through re-arrangement of debts, carry-on finance or concessional loans, who although having sound prospects of long term commercial viability, has used all his [sic] cash and credit resources and cannot meet his financial commitments.
- Farm Build-Up – To supplement through concessional interest rates and grants the normal processes under which properties which are too small to be economic are amalgamated with an adjoining holding or are subdivided and the subdivided portions are added to adjoining holdings, or to assist a farmer with a property too small to be economic to purchase additional land to build up his [sic] property to at least economic size.
- Rehabilitation – To provide limited assistance via a loan of up to $1,000 to those obliged to leave the industry where in the opinion of the Authority administering the scheme this is necessary to alleviate conditions of personal hardship. (derived from Commonwealth of Australia, 1971a)

While State authorities could divide the funding among the different components of the scheme as they saw fit, the Commonwealth recommended that half the assistance provided should be devoted to farm build-up purposes. This suggests that the main focus of the scheme was on building farms to a claimed viable capacity by amalgamating with those farms defined as unviable. Obviously, however, this depended on the willingness of neighboring farmers to sell their land.

It is important to note that the rationality underpinning the shift to 'reconstruction' was based not simply on objectives of improved efficiency: it was justified also in social terms. As the Schedule to the Bill states:

> The general principle to be applied is to distribute the available resources *as widely as practicable*, but the over-riding objective is to help to restore to economic viability those farm and farmers with the capacity to maintain viability once achieved (Commonwealth of Australia, 1971a:Schedule, Part I(b): italics V.H.).

This is a strange twist of a welfarist rationality of governing. While the objective of the legislation was welfarist in terms of ensuring that producers maintained their way of life through wide distribution of resources, this was applicable to only those who could meet the criteria for viability. In other words, farmers could pursue a 'way of life' in farming provided that they had prospects of maintaining profitability and low debt levels – given that debt had been defined by the BAE one of the key indicators of viability. This objective was emphasised by Minister Sinclair who stated: 'it is a policy of concentrating the assistance that is available on those cases which have the best hope of weathering the present adverse circumstances and recovering to a sound financial position in the longer term' (Commonwealth of Australia, 1971c:2334). Thus, the purpose of such assistance was to *restore* farmers' incomes to a level at which they could improve their productivity.

Despite its passage on the 25th of May 1971 into legislation, Opposition Labor members raised three criticisms during the Bill's Second Reading. The first was that 'viability', 'economic size' and 'sound financial position' were not defined in the legislation. In effect, this meant that the specific criteria in judging viability were left up to the judgment of individual State authorities who, while being able to assess past financial performance, had to make 'guesstimates' regarding future seasons and prices (Hefford, 1985). Obviously, this left much room for the State authorities in their assessment of applicants. This issue will be analyzed in more detail in the next section on State administration of the Rural Reconstruction Scheme. However, it is worth noting at this point that Labor members argued for more precise criteria through which long-term viability could be defined. For instance, the Labor member for Dawson, Rex Patterson stated:

> Who determines whether a producer has long term viability? I would like to know what assumptions are made… The Opposition wants to know more about eligibility because it is quite obvious from reading the schedule that there is not a sufficient definition of 'eligibility'. (Commonwealth of Australia, 1971c:2588)

Government members argued that assessments of viability and eligibility needed to be flexible. The Liberal Party's historical support of 'State rights', as opposed to the Labor Party's concern with centralization of government (see Jaensch, 1992), was here of significance in advocating a 'flexible' approach by the State authorities to assessment of applicants. Thus a more precise definition was not incorporated into the final legislation. Lack of a clear definition of viability and eligibility was claimed to be crucial given the States 'more complete knowledge of local conditions' and their closer proximity to the recipients of reconstruction assistance (Commonwealth of Australia, 1971c:2606). Nevertheless, some Labor members expressed reservations at giving the States such responsibilities in effectively dealing with a national problem. It was thus assumed that the States would operate within the broad guidelines of the Rural Reconstruction Scheme Schedule to achieve Commonwealth objectives. The problems associated with ensuring conformity with Federal guidelines at the same time as allowing State authorities scope for judgment will be discussed in more depth in the next three chapters.

A further criticism of the *Rural Reconstruction Bill* was that it encouraged a policy of 'get big or get out'. As Gilbert Duthie, Labor Member for Wilmot (Tasmania), noted: 'The scheme will make many big farmers bigger and will kill off thousands of small men [sic] who will be forced by banks, hire purchase firms and pastoral companies to sell out and get out' (Commonwealth of Australia, 1971c:2610). In other words, Duthie believed that the Scheme would encourage wealthy farmers and graziers to become richer through the purchase of neighboring properties, and the receipt of assistance via concessional interest rates. Certainly, this seemed to be encouraged through the farm build-up provisions.

Equally, while the RRS aimed to assist those farmers who sought voluntarily to leave agriculture, it did not compel unviable farmers to leave. It was more likely that those farmers defined as unviable would be forced off their land through foreclosure by banks rather than through any government program. If anything, the Scheme provided assistance for those farmers, defined otherwise as unviable by banks, to get through short-term difficulties and remain on their land. In any case, the drawing of causal links is a complex, if not impossible exercise for, as Musgrave (1990) argues, it is difficult to show that those farmers who leave agriculture through the assistance of adjustment programs would not leave anyway in their absence. That the 1971 Scheme encouraged farmers to 'get big or get out' is thus exaggerated. It represented more a technical means of defining the boundaries of viability through the constitution and categorization of the financial capacities of farmers. Its aim was to identify and classify farmers according to their capacities to achieve long-term financial viability.

A final criticism of the Rural Reconstruction Scheme, during Parliamentary debates, was that the rehabilitation loan was too small to be of use to farmers exiting the industry, and that the conditions to receive it were too strict. As Frederick Collard, Labor Member for Kalgoorlie (Western Australia), noted:

> Firstly, the amount of $1,000 is a ridiculously low amount and is unlikely to rehabilitate anyone. Secondly, unless the property which the farmer leaves is purchased under the conditions of the farm build-up section, he [sic] cannot qualify for assistance (Commonwealth of Australia, 1971c:2593).

Reviews of the Rural Reconstruction Scheme conducted in 1973 and 1974 show that rehabilitation was not popular among farmers. For instance, of all the potentially eligible producers, just over two per cent had by March 1973 sought rehabilitation assistance (BAE, 1973). Nevertheless, from the Commonwealth government's perspective in 1971, this provision constituted an important part of the Scheme due to its objective of assisting farmers for welfare reasons when financial conditions became such that exit from the industry was the only option.

Following joint Federal and State Government reviews between 1972 and 1974 the Rural Reconstruction Scheme was extended in a number of ways. Many of these changes were initiated under the new Whitlam Labor Government, elected in December 1972. Changes are outlined below.

- The introduction of a Rural Reconstruction Employment Training Scheme designed for those farmers likely to be refused debt reconstruction assistance, and those whose properties were acquired under the farm build-up provisions. Due to low numbers of applicants this scheme was amalgamated with other schemes in 1973.

- Agreement by the Commonwealth to spend $100 million over two years instead of four, and provision of an additional $15 million due to high demand for assistance in the Scheme's first year.
- Continuation of rehabilitation loans, despite their poor uptake, and a raising of their upper limit to $3,000 in 1972. In 1974 it was agreed that the States should be free to decide whether such assistance should be in the form of loans or grants.
- Approval at the second review in 1973 for the Scheme to run for an additional three years.
- Enhanced emphasis on farm build-up through the formulation of formal objectives between the Commonwealth and States in 1973.
- Special arrangements for fruit-growers under the 'tree-pull' scheme when their markets collapsed as a result of Britain in 1973 entering the European Economic Community. (derived from BAE, 1974 and Hefford, 1985).

The changes noted above were crucial in constituting 'reconstruction' as an ongoing problem that needed to be better governed in order to achieve national objectives of a more productive and market-responsive agriculture. This was given added emphasis when in 1974 the Industries Assistance Commission was created, a point discussed in detail in Chapter Two.

GOVERNING 'LOW-INCOME' FARMS: ADMINISTRATION BY THE STATES

The Australian States became involved initially in the governing of low-income farmers during 1970 with the passage of the Marginal Dairy Farms Reconstruction Scheme. However, it was not until the introduction of the Rural Reconstruction Scheme in 1971 that the States agreed to set up statutory authorities to deal directly and specifically with reconstruction issues. In many cases, this involved extending the functions of existing land settlement or rural finance authorities although, in Queensland and Western Australia where such institutions either did not exist or could not be adapted to different roles, new administrative bodies were established (BAE, 1973). Once in place, it was then up to the individual State authorities to administer assistance consistent with the broad objectives of the Commonwealth government. In line with the 1971 legislation, the Commonwealth government recommended that the States devote half the assistance to farm build-up purposes. According to the Schedule, this involved calling for applications for assistance, determining eligibility, approving or rejecting applications, determining the amount of assistance for successful loan applications and deciding on the terms and conditions for such loans.

The Liberal/Country Party Coalition government had argued in 1971 that the considerable administrative flexibility left to the States was important in ensuring 'that each authority would not be restricted in tailoring its procedures to meet the different circumstances and needs of producers and industries in each State' (BAE, 1973:2). Unfortunately, such 'flexibility' also meant substantial variations between States in terms of different operational objectives, guidelines and procedures for determining viability and thereby eligibility. For instance, in South Australia on the one hand – where the Labor Party led by Don Dunstan was in government – applicants had to supply five years of historical data and full budget estimates of production, livestock trading, income and expenditures for the current year's operations. On the other hand, in New South Wales – where a Liberal/Country Party was in government – applicants needed only to provide evidence of their farm's operations in a 'normal' year (BAE, 1973:9). The differences in assessment between the two States are therefore considerable. They provide an insight into the various interpretations of Commonwealth objectives among the States, and the opposition, particularly by Liberal/Country Party States, to the Whitlam Labor government's desire to achieve a diminished role for State governments.

Moreover, each of the States placed a different level of emphasis on 'measures' of viability such as debt/asset, returns/cost and surplus/debt ratios. Clearly these differing procedures of assessment, and methods of determining eligibility, demonstrate the variation in how viability was defined, and particularly what was regarded as 'commercial viability' or a farm of 'economic size'. Farm size alone was not seen as a key factor in determining eligibility. This was despite claims by the BAE that productivity could be improved through increases in farm size. Farmers' levels of equity and indebtedness were found to have the most significant bearing on the decisions made by State authorities in deciding on long-term viability, and thus eligibility (BAE, 1973). Nevertheless, the general assessment procedures and methods of determining viability showed much variation. As a consequence, a unified representation of the characteristics that constituted a viable farm was very difficult to achieve in practice.

The debate concerning the capacity of State authorities to follow, in a coordinated matter, Commonwealth objectives illustrates a central problematic of governing 'at a distance': how does one ensure, in the absence of direct monitoring, that actors will conduct themselves in the desired manner? As the Commonwealth depends on the States to administer assistance to farmers under section 96 of the Constitution, guidelines need to be flexible. For example, attempts by the Whitlam Labor government between 1972 and 1975 to adopt a more heavy-handed and centralized approach to administration under the banner of the 'new federalism' – resulted in widespread opposition by State governments, particularly in New South Wales, Queensland and Victoria where conservative governments held office. An important exception to this was the Dunstan Labor government in South Australia, which embraced enthusiastically greater federal powers (see Jaensch, 1992:86). Such resistance draws attention to the need for forms of rule that enable flexibility, but that also achieve national objectives. It is difficult to make clear causal links between the procedures and practices of individual State authorities, and the number of applicants or rate of approvals. This would involve much more detailed work on the processes of individual authorities as they applied to specific cases, and the diversity of factors influencing farmers to apply for assistance in the first place, both of which are beyond the scope of this book. However, some general points can be made concerning the nature and extent of assistance provided, and its consistency with Commonwealth objectives.

While the Commonwealth sought to promote the farm build-up component of the Rural Reconstruction Scheme, over two-thirds of all approved applications up to the end of June 1974 were for debt reconstruction (BAE, 1974). The percentage of total applications for debt reconstruction was even higher. This would seem to indicate three things. First, that debt reconstruction suited the financial circumstances of farmers at this time. Certainly the downturn in the wool industry during 1970/71 would have had a considerable impact on farmers' levels of credit. Second, varieties of debt reconstruction assistance had been in existence since 1935 and the State authorities were more accustomed to administering this as opposed to other forms of support, despite increased levels of funding being allocated to farm build-up. Third, debt reconstruction assistance accorded more closely with welfarist rationalities of governing, and providing assistance to maintain farmers on their land. That is to say, it did not involve property amalgamation and therefore a displacement of farmers.

The consistency between the actual number of applications and the approvals would seem to indicate a combination of these three factors. In any case, as noted above, formal objectives to enhance the emphasis on farm build-up were made between the Commonwealth and the States in 1973. This served to slow the rate of approvals for debt reconstruction in 1974, and increase those for farm build-up purposes (BAE, 1974). The number of applicants for rehabilitation assistance remained very low, although evidence from the BAE suggested that the approval rate was not dissimilar from the other components of the Rural Reconstruction Scheme.

Rural Reconstruction in Queensland

In Queensland, similar legislation to the Rural Reconstruction Act had been in operation since 1967 under the *Farmers' Assistance (Debts Adjustment) Act* (Parliament of Queensland, 1967a). This legislation was enacted under the conservative Nicklin Country/Liberal Party government at a time when many producers were suffering from a prolonged period of low rainfall and rising debt levels. Its purpose, according to Gordon Chalk – Treasurer and Country Party Member for Lockyer (a farming district) – was to keep as many farmers as possible engaged actively in the industry and to 'overcome their difficulties as time goes on' (Parliament of Queensland, 1967b:2752).

The legislation therefore had relatively clear social objectives: its purpose was to keep farmers on the land, provided that they were not completely insolvent. This rationality of governing was shared by both parties, as the following statement from John Houston, Leader of the Labor Party, demonstrated:

> Queensland is still basically a primary-producing State; to a great extent our economy is tied to the success of our primary industries, and it is likely to be for a long time. Apart from the humane angle – from a purely economic angle – we cannot allow those in our primary industries to be lost to us. (Parliament of Queensland, 1967b:2733)

Viability in this context was taken to mean a rearrangement of debt commitments so that a producer could remain on their land. Assistance to farmers to maintain them on the land was the rationality for governing farmers, as it was in 1967 at the Federal level of government.

In anticipation of the passage of Commonwealth rural reconstruction legislation, the financial situation of farmers came to be increasingly seen in Queensland politics as a broader economic problem. Not only was the debt situation of farmers now framed as a problem of national significance, it was seen also to require 'reconstruction' measures, not simply debt adjustment. As Victor Sullivan, Minister for Lands, and Country Party Member for Condamine (another farming district), noted, 'Of course, debt adjustment is something quite distinct from farm reconstruction in the broad context. Total reconstruction denotes a very comprehensive action plan' (Parliament of Queensland, 1970:2171). It is worthwhile noting here how reconstruction was differentiated from debt adjustment and represented as a broader, and thereby more serious, problem. The focus on reconstruction rather than debt adjustment was manifested materially with the establishment on the 16th of December 1970 of a Rural Reconstruction Authority comprised of public servants.

The Rural Reconstruction Board was created through an amendment to the *Farmers' Assistance (Debts Adjustment) Act* (1967), and its primary task was to administer the assistance under this Act. Only after the Commonwealth *Rural Reconstruction Act* was passed in May 1971 did the Board take responsibility for administration of this legislation. Further amendments were made to the Act in 1971 to provide for the establishment of a Rural Reconstruction Fund and a Woolgrowers Assistance Fund, and to empower the Board to administer monies provided by the Commonwealth and/or State for these purposes (Rural Reconstruction Board, 1972). Unlike its predecessor, the Agricultural Bank, which simply administered assistance, one of the first functions of the Rural Reconstruction Board was to inquire into the problems affecting the rural industries and suggest remedial measures for consideration by the State Government (Parliament of Queensland, 1970:2172). It therefore played an important role in both making knowable reconstruction as a distinct political and administrative problem, and in constituting the strategies that the Government might pursue to address such a problem.

The establishment of the Rural Reconstruction Board represented a distinct break with previous rationalities for governing farmers in Queensland. Not only were agricultural economists accorded a more prominent role in the assessment of viability, but farm business management began to emerge as a legitimate means of 'reconstructing' farms. Farm management was to have a more central influence in the governing of agricultural change in later years. The publication in 1970 of a book entitled *Farm Management and Agricultural Economics* (Hardaker, Lewis and McFarlane, 1970) provides a useful indication of the influence that agricultural economists were beginning to have in the emergence of a distinct discipline of farm planning. According to Michael Ahern, Country Party Member for Landsborough (a farming district), such rational planning was essential to the operations of the new Rural Reconstruction authority:

> It is very easy to 'rubbish' farm business planning as a discipline, but it is essential for the success of Queensland's Rural Reconstruction Board, and it is absolutely essential also that new economic techniques and adequate budgets and plans be provided for the properties assisted (Parliament of Queensland, 1970:2450).

Despite the enhanced expertise accorded to economic disciplines in addressing the emerging problem of low-income farms, a welfarist rationality of governing remained prominent. Not only do Parliamentary debates of the period show that rural reconstruction assistance was important in keeping farmers on their properties until weather patterns and commodity prices improved, but the subsequent decisions made by the Board show continued support for debt reconstruction (see Rural Reconstruction Board, 1972). While the Commonwealth-State agreement in 1973 re-focused State efforts toward farm build-up, a welfarist rationality of governing continued to underpin assessments, even though the specific technologies applied for defining farm eligibility, such as budget planning and forecasts of financial performance, might have contradicted these objectives.

FARM ORGANIZATIONS AND COUNTRY PARTY PATRONAGE

From the 1950s until 1973, the governing of farmers and graziers was underpinned by a welfarist rationality of rule. This type of rationality was based on the belief that government assistance – through subsidization, price support, and other forms of risk collectivization – would maintain the viability of farmers on their land. As the only political party that appeared to have a concern with the interests of Australian farm families, the Country Party was significant during its 23 years in office with the Liberal Party in maintaining this rationality of governing, and in defining the farm policy agenda (Hefford, 1985:64). However, in defining farm policy, the Country Party invited farm organizations to play a central role. This was illustrated in a comment made in 1965 by the leader of the Country Party, John McEwen:

> My attitude is that neither the Australian Country Party nor its parliamentary members should decide what is the correct policy for a primary industry. It has always been the policy of my party that those who produce, own and sell a product are the best judges of the way in which their own property should be treated. It is the function of my Party to see that the will of those who produce and own the product is carried into legislative and administrative effect. (cited in Hefford, 1985:65)

This statement demonstrated that the Country Party was more than willing to ensure that the interests of producers, expressed through their representative organizations, were effected into policy. Many farm organizations were, in fact, accorded virtually unrestricted access to Country Party Ministers. It soon became 'conventional wisdom' that all farmer organizations needed to do in order for a particular policy to be implemented was to have an informal conversation with Country Party members (Trebeck, 1990:134). Equally, as Connors (1996) notes, the career path for compliant farm leaders was well known and very attractive: 'There were seats on, and the chairmanship of, marketing boards and other statutory authorities, places on overseas trade missions, the party's blessing for parliamentary pre-selection, and imperial honours, including knighthoods' (172).

Nevertheless, this political patronage had its limits. In practice, only the interests of some groups were accorded priority. The Country Party's platform of 'protection all round', particularly under the leadership of John McEwen, sought to obtain concessions for small commodity producers who, it was believed, would be unable to remain economically viable without government assistance. Those commodity groups such as the Australian Woolgrowers and Graziers' Council, who pursued actively a more open market, opposed the Country Party's commitment to protectionist policies, and particularly their support of tariffs (Connors, 1996). This reflects, in fact, the historical division in Australia between small commodity producers, who generally support the extensive government intervention favoured by the Country Party, and the larger graziers and woolgrowers, who endorse freer trade and minimal government interference (Encel, 1970:314).

While John McEwen remained leader of the Country Party, the interests of smaller commodity producers were frequently represented as those of all primary producers. It is perhaps not surprising then that between the 1950s and 1972 farmers' incomes were protected through subsidies, tariffs, bounties and other similar forms of assistance designed to avert the risks of fluctuations in demand. By the end of the 1960s, however, electoral support for the Country Party was waning and, as discussed in a previous section of this chapter, the Liberal Party issued its own statement of objectives for farm policy. The decline in electoral support, and the increasing scepticism among Country Party members (such as Doug Anthony) that previous strategies of price support had not worked, enabled the anti-protectionist interests of larger woolgrowers to assume greater political prominence.

This was facilitated further through the election of the Labor Party to government in December 1972. The Labor Party had opposed vigorously the level of patronage afforded by the Country Party to particular farm organizations. Their election changed fundamentally the way in which farm organizations interacted with government. On being elected, the new Minister for Primary Industry, Senator Ken Wriedt, foreshadowed change when he stated that farm policy should be based on rational management rather than the right of producers to assistance (Marshall, 1986:29). As noted in the next chapter, this was reflected in the establishment, in 1974, of the Industries Assistance Commission (IAC). Rather than gaining direct access to politicians, farm organizations, in order to remain relevant, now had to become more 'professional' and to formulate written submissions. This tended to provide an advantage to the larger woolgrowers who saw the IAC as an state-sanctioned body that could legitimize their belief that many farmers were being insulated from greater competition through government assistance (Connors, 1996:175).

Moreover, grazier groups were more familiar with the economic expertise now required to make proper submissions. These groups had used economists to argue for the removal of protection, and the pursuit of freer trade. Thus, grazier groups, with interests primarily in wool and beef, tended to oppose the IAC far less than those farm organizations representing smaller commodity producers involved in such industries as dairying and wheat. At the same time, however, the level of resources required to make submissions for the IAC increased calls by producer groups for greater unity among farm organizations. The establishment of the National Farmers' Federation in 1979 (discussed in more detail in Chapter Two) eroded further the rationality of agricultural regulation through welfarist forms of risk collectivization.

CONCLUSION

The governing of agricultural change through a welfarist rationality of rule was called into question, or problematized, from the late 1960s. As I argued in this chapter, a 'low income farm problem' was produced through the knowledge, techniques and calculations of agricultural economists as a form of expertise. The activities of economists in seeking to describe change were crucial in making visible the practices of farmers, and in representing previous forms of governing as inadequate in contributing to national productivity and farmer welfare. In fact, economic expertise was central in representing in a knowable form, the existence of low-income farms, and in constituting the characteristics of viable farmers. The greater role given to the BAE by the Country Party Minister for Primary Industry, Doug Anthony, and the exit of John McEwen as leader, was important in enabling low-income farms to become an economic issue worthy of long-term attention by both governments and farm organizations.

In categorizing farmers on the basis of income, debt levels and equity, a strategy of 'reconstruction' emerged as a means of responding to the low-income problem constructed by economists. Reconstruction comprised programs of farm build-up, debt adjustment and rehabilitation, the objectives of which were to improve farm sector viability. However, while there remained a desire to govern in a 'social' way, the greater role accorded to economic expertise questioned welfarist technologies of governing that had collectivized the risk of farmers. A tension emerged between objectives of maintaining producers on their land through 'protectionist' forms of assistance, and the desire to improve farm incomes through reconstruction measures. The effect of this tension was a focus by State authorities on improving incomes and achieving viability through more traditional forms of assistance such as debt adjustment. While the encouragement of farm build-up by the Commonwealth, from 1972, sought to concentrate attention on those farmers deemed to have the capacities to improve their productivity in the longer-term, the 'flexibility' accorded to State authorities resulted in much variation in assessment procedures, and thus considerable differences in definitions of eligibility and viability.

The next chapter examines how the period 1973 to 1985 was characterized by a concern with governing agricultural change through better economic management techniques. Establishment of the IAC in 1974 represented a decisive shift in the evaluation of agricultural assistance, and entrenched the authority of economic expertise in assessing the efficiency or otherwise of restructuring measures. However, the increased authority of economic knowledge was shaped by a continuing concern with distributing resources as widely as practicable, particularly at a State level.

Chapter 2

A 'RATIONAL' APPROACH TO ECONOMIC MANAGEMENT: LOW-INCOME FARMS AS A LONG-TERM EFFICIENCY ISSUE

As argued in the previous chapter, a 'low-income farm problem' was identified by agricultural economists from the late-1960s as an issue of national *economic* significance. Previous techniques of managing agricultural change were called into question as a consequence, and new programs such as the Rural Reconstruction Scheme introduced to ensure that adjustment to change occurred in a way that contributed to farm economic viability and the welfare of farmers. Nevertheless, as I argued in Chapter One, the administration of assistance remained underpinned by a welfarist rationality of governing. This chapter draws attention to the ways in which a welfarist way of reflecting on and conducting rule was problematized further through the 1970s and early-1980s. Of particular interest in such problematizations of rule is how practices of regulating low-income farms posed a major problem for long-term national efficiency if not governed in a more economically 'rational' manner.

The 1970s in Australia saw a shift in political thought concerning the economy and the appropriate role for governments in economic regulation. According to Head (1989:498), this shift was based on the assumption that first, the international political economy had become fluid, unpredictable and threatening; and second, that the institutional structures and policy positions of the boom years before 1974 were inadequate for dealing with such conditions. From the perspective of an analytics of government, a welfarist rationality of governing was subject to increasing question, and seen by many authorities as outdated in terms of improving overall economic efficiency in agriculture. National growth was argued by governments to be possible through improved management of the economy as a whole. In fact, apparent success in economic management became an important strategy during this period through which governments could maintain control of the political agenda (Head, 1989:486).

The creation of the Industries Assistance Commission (IAC) in 1974 represented a key part of this shift in rationalities of governing emphasizing, as part of its objectives, the need to evaluate policy options with reference to general economic management (see Martin, 1990). As this chapter argues, the creation of the IAC was of particular significance in enhancing the role of economists in the public service, and transforming economic expertise into an obligatory form of advice in the preparation of 'professional' submissions. It also increased the need for economists in farm organizations, who were expected to evaluate their requests for assistance according to economic criteria. Nevertheless, this chapter argues that an uneasy tension existed between attempts to manage low-income farms as a problem of long-term economic efficiency, and the rationality of the National Country Party[1] (in government with the Liberal Party between 1975 and 1982), and many of the State authorities, to recognise the social importance of maintaining farmers on their land.

Low-Income Farms as a General Economic Management Problem

The election of a Labor Government in December 1972, led by Gough Whitlam, resulted in a significant shift from previous approaches to governing. This was the first Australian Labor government in 23 years, and it sought to introduce a new system of economic management focused on deregulation and competition. It vehemently opposed the system of agricultural regulation – such as subsidization, tariff and price support – that had been established by the previous conservative Liberal/Country Party coalition governments. These forms of regulation, described by Butlin, Barnard and Pincus (1982) as comprising 'all the restrictive practices known to man [sic]', were believed by the Labor Party to result from a sectional concern by the Country Party with maintaining the economic and social position of their farming constituents.

The Labor Party opposed the patronage that it observed between the Country Party and farming organizations on the grounds that it had little regard for the consequences of protectionist policies on taxpayers and consumers (Watson, 1979:165). One of the new government's first steps was therefore a 25 per cent across-the-board removal of tariffs in July 1973 (Catley, 1996:61-62). This was followed in February 1974 by the removal of the bounty on superphosphate. However, the most important change that sought to challenge the Country Party's political dominance in rural matters was the establishment in 1974 of the Industries Assistance Commission (IAC).

The creation of the IAC was crucial in changing the way that industry policies were made and evaluated. Increasingly, policy-making began to be dominated by formalized economic approaches, with government assistance assessed in terms of its effects on economic efficiency (Martin, 1990:156). Rather than speaking directly to the relevant Minister, producer groups now had to prepare detailed submissions. According to Martin (1990), the establishment of the IAC represented 'the most important single institutional change affecting agricultural policy during the 1970-83 period... Its introduction changed the rules of the policy formulation game' (157). Not only did the creation of the IAC open to more critical scrutiny the processes by which assistance to farmers was allocated, it also enabled economic expertise to provide high levels of input into how these processes might most efficiently be governed in relation to the economy *as a whole*. Thus, as Edwards and Watson (1978:226)

[1] Between 1975 and 1982, the National Party was called the National Country Party. Prior to 1975 it was known as simply the Country Party. From 1983 it became the National Party.

note, the creation of the IAC enhanced considerably the status and influence of agricultural economists both in the public service, and in academic institutions.

The economic arguments and skills required in the process of making formal submissions gave economists and, more specifically, agricultural economists a distinct advantage over traditional administrators that had predominantly science backgrounds. Economic expertise was now essential in transforming the political concerns of both farm organizations (as representatives of farmers), and governments into, what Rose (1993) calls, a 'non-political' discourse of truth. Those that based their claims for assistance on sectional (political) interests would henceforth find their claims de-legitimized. Obviously, this changed the way in which farm organizations operated; a point that will be developed further in a later section of this chapter. The objectives underpinning the creation of the IAC will now be outlined, and their significance to the governing of agricultural change examined.

Creation of the Industries Assistance Commission (1974)

The Industries Assistance Commission Bill was introduced into the Federal Parliament in September 1973. Its main aim was to extend the advisory functions of the Tariff Board beyond manufacturing to encompass other industries. Essentially, the role of the Tariff Board was to make public inquiries and reports on questions of assistance relevant to manufacturing referred to it by government. According to the Labor Prime Minister, Gough Whitlam, an extension of this process to all Australian industries would not only contribute to more efficient use of the nation's resources, but also 'substitute a deliberate, systematic, and comprehensive program of public inquiries for the rather random, haphazard, and sometimes informal and superficial processes of investigation of the past' (Commonwealth of Australia, 1973a:1633). In other words, Whitlam believed that a more systematic and 'rational' program of public inquiry applied to all industries would remove the political patronage and appeal to sectional interests of the past, and allow the introduction of a more publicly transparent and 'non-political' inquiry process. This transparency was to be achieved through the formal process of submissions. One can see here the emergence of what Dean (1999:168-169) calls 'technologies of performance' into formal political practice in which economic techniques of calculation exercise seemingly more transparent and truthful scrutiny over the conduct of authority. The use of formal submissions based on economic expertise was, in fact, a powerful means of exercising increased surveillance over the conduct of policy-making – particularly the conduct of Country Party members.

The purpose of the IAC, however, was not just to improve transparency. Its deliberations were to be conducted within a framework of general economic management (Commonwealth of Australia, 1973a:1633). This represents a significant shift away from the objectives of the Tariff Board in that the IAC's recommendations on particular issues had to be framed in terms of their effects on the prudent management of the economy as an interrelated whole. That is to say, the efficiency of the economy in its entirety was to be the primary factor in considering what recommendations should be made for certain industries, rather than the economic welfare of specific industrial sectors. This called into question a welfarist rationality of governing and the overall economic effects of attempts by authorities to collectivize the risk of specific industries. The IAC was established to achieve economic growth, not through socialized means, as was the case with the Tariff Board, but to evaluate these earlier forms of intervention in terms of their role in contributing to efficient economic management.

The distinctly economic agenda of the IAC can be seen through its objectives. These were to:

(a) improve the efficiency with which the community's productive resources are used;
(b) encourage those economic activities in Australia, and the producers of the goods and services concerned, which contribute to improving the efficiency with which the community's productive resources are used;
(c) facilitate adjustment to changes in the economic environment by industries and persons affected by those changes;
(d) recognize the interests of consumers and consuming industries likely to be affected by measures proposed by the Commission;
(e) ensure that any measures for assistance to, and development of, industries are integrated with national economic policy as a whole;
(f) ensure that Australia's trade and protection policies are compatible; and,
(g) provide adequate scope for public scrutiny and evaluation of the basis of the Commission's reports (Commonwealth of Australia, 1973b:s.22(1)).

The IAC's establishment, while supported by the conservative Liberal Party, was opposed vehemently by their Coalition partners the Country Party, whose electoral support was based in country areas. The Country Party argued that the purpose of the IAC's creation was for the Labor Party to legitimize a withdrawal of assistance from rural industries and communities. As many of the Country Party's constituents depended on government support, this was, perhaps, a genuine political concern. However, the Country Party's main concern with the legislation was the de-legitimation of previous practice where industry leaders negotiated directly with government ministers concerning assistance levels. With the passing of the IAC legislation, representatives of farming industries had to prepare detailed reports that accounted for how the assistance measures they suggested would contribute to not only the economic efficiency of the industry concerned, but also to the general management of the economy as a whole. As the recently-appointed leader of the Australian Country Party, Doug Anthony, stated during the second reading of the IAC Bill:

> ...industries will find themselves talking to wooden-faced bureaucrats instead of to Ministers who at least can be responsive to the approaches of industry and who can apply – as should be applied – other than the cold, hard economic criteria on which decisions will be made by bureaucrats who have no direct responsibility to the people (Commonwealth of Australia, 1973a:2357).

From the governing Labor Party's point of view, however, such sentiments were reflective of the sectional interests that the Country Party attempted to please when it was in government with the Liberal Party. The Labor Party's argument was that the adoption of a more rational and systematic approach had the capacity to treat all interest groups equally and was therefore far more transparent to the public than the hidden arrangements between Country Party Ministers and farm organization representatives. According to Bill Hayden, Labor Member for Oxley (Queensland) during the second reading of the Bill:

> ...if unwise decisions are being made not according to rational economic criteria but according to the pay-off that can be achieved in the back room deals between the Country Party and the people who are its backers...we will not achieve the sort of growth we should achieve (Commonwealth of Australia, 1973a:2372).

Such arguments attempted to constitute the formal procedures of the IAC as a rational and more truthful means of evaluating the claims of different groups as opposed to past practices. In this respect, the IAC was constructed as the only way of achieving public transparency and truthful outcomes in evaluating industry assistance.

The Liberal Party agreed also that a more rational approach to policy-making was essential. As Bill Snedden, Member for Bruce, argued:

> The economy, of course, must be seen as a whole in a modern economy. The different sectors are so closely linked that that we could not afford to concentrate on one sector to the exclusion of all else (Commonwealth of Australia, 1973a:2429).

Statements such as this represented a shift of emphasis away from agriculture as the key to Australia's growth, towards a model of economy in which all industries were constructed as competing on a level playing field. In sharp contrast, the *Green Paper on Rural Policy*, published in 1974, sought to contest the rationality of governing within the IAC legislation, and assert the significance of a welfarist governmental rationality in regulating agricultural change.

Green Paper on Rural Policy (1974)

At the same time that the IAC legislation was being debated in Federal Parliament, the Whitlam Labor government convened a working group, comprised of public servants, to prepare a Green Paper on Australian rural policies. The establishment of the working group represented a response, in part, to the adverse reactions by farmers to the 1973 Federal budget that had eliminated a number of farm assistance measures, including dairy subsidies and the superphosphate bounty (Watson, 1979:166). Published in 1974, and subtitled *The Principles of Rural Policy in Australia* (Harris, Crawford, Gruen and Honan, 1974), the *Green Paper on Rural Policy* was essentially an outline of principles to guide rural policy-making, rather than a statement of government policy. While it did not become a White Paper, Campbell (1974) argues that it was, nevertheless, an important document in that it attempted to draw together much of the 'conventional wisdom' of the agricultural economics profession in Australia as it had developed in the previous 25 years.

The Green Paper's central focus was on how governments could intervene in order to improve the workings of the market. While the market was seen as the 'most effective way of allocating resources in the agricultural sector' (Harris et al., 1974:32), the Green Paper's authors concurred with Harris (1970) that government intervention was necessary to both improve the operations of the market and to compensate for its harsh consequences. The working group argued that market forces left to themselves were damaging to both economic efficiency and the welfare of farmers. As Harris et al (1974:48) state:

> A national rural policy should be an integral part of a national economic policy and should be judged in terms of its contribution to national economic and social objectives with equity not only between those in rural areas and those in the rest of the community, but also among those in rural areas.

This statement, while arguing that agriculture should no longer be a special case in terms of allocating assistance, sought to constitute the terrain of governing in terms of a distinctively welfarist political rationality. That is to say, both social and economic objectives were constructed as mutually reinforcing goals (see Dean, 1999) with economic equity argued to be the *social* responsibility of governments.

In order to achieve economic objectives in a social way, tariff compensation was suggested as the best strategy. Tariff compensation involves compensating those export industries with low levels of protection in order to make up for the higher prices imposed on these sectors by tariff protection in other industries, such as in manufacturing. The objective of a tariff compensation argument was a more equitable distribution of resources based on the assumption that goals of economic efficiency and equity of income among farmers were mutually achievable. However, the argument for tariff compensation by the working group was unique among government-sponsored documents of this time (Campbell, 1974). While being welcomed by farming organizations, many economists opposed tariff compensation on the grounds that it would result in further distortions in the economy and involve inefficient use of resources. The IAC, in particular, was extremely critical in its first Annual Report of the tariff compensation approach, arguing that it involved costs to Treasury and that it conflicted with Australia's overseas trade interests (Watson, 1979:166).

While tariff compensation confirmed the Green Paper working group's desire for social forms of governing, their arguments concerning how best to manage agricultural change show some ambivalence. A tension is evident between the desire to govern in a social way, and the acknowledgment that this form of governing might have been impeding economic efficiency by keeping so-called 'low-income' farmers on their land. For instance, the Green Paper states that government assistance should 'aim to assist rather than impede market forces' (Harris et al., 1974:196). Nevertheless, the working group recognised also that the existence of a low-income problem was often a result of market failure. They argued, therefore, that the proper task of government was in certain cases to 'moderate the influence of market forces' in the interests of social welfare and efficiency and to ensure that adjustment was not enforced, but occurred 'in an orderly manner' (Harris et al., 1974:195-196). In other words, while conformity to market demands was seen as desirable, certain special conditions, such as a high proportion of low-income farms in a rural industry, warranted government intervention in order to mitigate social welfare problems, and to ensure an efficient allocation of resources. On this basis, the working group called for the continuation of the Rural Reconstruction Scheme along more flexible lines that assisted farmers disadvantaged by sudden changes in economic conditions.

In many ways the Green Paper represented an attempt to constitute the rural sector, and attempts to govern it, in a distinctively welfarist way. However, the Green Paper's difference from previous reports was its explicit reference to the role of government in assisting market forces. The working group argued that there were only *some* circumstances where a role for governments was justified on efficiency and equity grounds in moderating the impact of market forces. This is different to rationalities of governing between the 1940s and 1960s where market failure was believed to always warrant government intervention (Edwards and Watson, 1978:200; see also Butlin et al., 1982). While the working group's tariff compensation argument may not have been consistent with the objectives of the IAC, its questioning of 'protection all round', and its attempts to evaluate assistance measures in an economy-wide context, showed some similarities. The demise of the Whitlam Labor government in 1975, and its replacement with a conservative Liberal/National Country Party Coalition government led by grazier Malcolm Fraser served to refocus political attention on how assistance to farmers should best be evaluated.

Election of a Liberal/National Country Party Coalition Government (1975)

The ambitious reform program of the Whitlam Labor government had been pursued vigorously during 1973. However, a world economic downturn in 1974, the entry of Britain into the European Economic Community, and the OPEC-induced increases in oil prices of the same year, resulted in the most severe economic recession in Australia since World War Two – thus putting an end to these ambitions (Catley, 1996:62). The election of a Liberal/National Country Party Coalition government in December 1975, following the historic dismissal by the Governor General of the Labor government in November, resulted in a renewed focus on formal approaches to the evaluation of agricultural assistance. Nevertheless, many of the assistance measures abolished during the period of Labor rule were restored, such as price support arrangements and tariff levels (Watson, 1979:168). Such measures were aimed primarily at placating farmers who had protested strongly against the Whitlam government's attempts at agricultural reform. As the source of the National Country Party's electoral support was based in rural electorates, the Coalition government could not afford to ignore calls from farm organizations for support to struggling producers.

The economic viability of low-income farmers was seen as an important priority for the new government with the IAC asked to report on the types of measures that should be put in place following the expiry of the Rural Reconstruction Scheme in mid-1976 (Burdon, 1996:14). It is important to note that the inquiry into rural reconstruction assistance was commissioned originally in 1974 by the Whitlam Labor Government, but was not actually completed until 1976 by which time the conservative Liberal/National Country Party coalition government was in office. The inquiry was conducted by the Industries Assistance Commission (IAC) who were asked to report by the Commonwealth Department of Primary Industry on what measures should be put in place after the expiry of the Rural Reconstruction Scheme in mid-1976. Commissioning the IAC to report on rural reconstruction measures represented a notable political move in that part of that organization's role was to give closer scrutiny to established assistance measures, and to consider the effects of particular policies in an economy-wide context. Thus, a tension is evident between the IAC's objectives of contributing to economically efficient policy outcomes, and the Fraser government's desire to restore assistance measures removed under Labor rule. This tension is evident in the parliamentary debate and legislation that followed from the IAC report, discussed in detail below.

IAC Report on Rural Reconstruction (January 1976)

The IAC framed its report on rural reconstruction in terms of the international economic changes that had taken place during 1974-1975. For instance, it emphasized in its report that while the entry of the United Kingdom into the European Economic Community had resulted in greater national dependence on other markets, increased wealth in oil-producing nations had opened new export opportunities. The IAC argued that such changes had the potential to undermine prices in a sector reliant on exports, and increased considerably the pressures for structural change in Australian agriculture. Farmers earning incomes 'low in relation to incomes earned in other locations and occupations' (IAC, 1976:14) were therefore seen by the IAC as a problem that needed to be addressed in order for prices received by farmers to remain stable. The presence of a claimed low-income farm problem, however, was not of itself considered to provide sufficient grounds for government intervention. Instead, the IAC (1976:14) argued that, 'normal forces of competition continually create pressure for structural change to which the community adapts without economic cost or hardship'. A role for government was justified therefore on the basis of 'smoothing' the process of change, ensuring that available resources were used efficiently and that hardship on individuals was minimised (see also Kingma and Samuel, 1977). In fact, government inaction was seen by the IAC to *damage seriously* the confidence of farmers and lead to unwise responses in their management practices.

The IAC reviewed existing reconstruction programs including, the Rural Reconstruction Scheme, the Dairy Adjustment Program and the Fruitgrowing Reconstruction Scheme and found that these schemes had not provided the necessary opportunities for farmers to adjust to demand-based prices. In reviewing these programs, the IAC raised some important policy issues including:

- the extent to which Government funded reconstruction schemes were justified on the grounds of efficiency and/or welfare;
- whether one general scheme was better than a number of industry-specific schemes;
- the range of measures needed to achieve the Government's objectives;
- the eligibility criteria used in allocating funds;
- the terms and conditions attached to the provision of funds;
- how the scheme(s) could best be administered;
- the degree of flexibility and discretion exercised by the administering bodies; and,
- the details of the Federal/State agreement' (Industries Assistance Commission, 1976:39-40).

The IAC focused particularly on the Rural Reconstruction Scheme and recommended that its principles and policy instruments be extended into a new program integrating all previous adjustment measures: this was to be called the Rural Adjustment Scheme.

The change in language from 'reconstruction' to 'adjustment' represented a key shift in governmental rationalities. While reconstruction programs comprised measures to alleviate the impacts of events constructed as part of short-term downturns, 'adjustment' was constituted as a long-term means of responding to problems seen as *structural* in nature. The term adjustment had been in usage within various agricultural policies since the 1930s, primarily in relation to the short-term restructuring of debt. However, its emergence in the IAC's 1976 review redefined adjustment as an ongoing element of *general economic management* in which Australian agriculture could be 'adjusted' to claimed economic realities. Thus, as the IAC (1976:74) noted, 'Government assistance for farm adjustment should be recognised as a continuous and ongoing program assisting the rural sector to *anticipate and respond to* changes in production patterns' (italics V.H.). The IAC's objective of examining issues in terms of their effects on national economic efficiency was now significant in constituting low-income farms as an adjustment problem that governments should govern in the interests of prudent economic management. In other words, as a statutory form of expertise, the IAC played a central role in constructing low-income farmers as a distinct group whose capacities, with the help of government intervention, could be shaped to improve the workings of a national economy.

A number of other key changes were recommended also by the IAC. These included:

- A strengthening of the commercial emphasis of rural adjustment assistance.
- An extension of the farm build-up provisions to include the acquisition or build-up of any non-land resource considered necessary for the achievement of long term commercial viability of the farm enterprise.
- The provision of short term carry-on funds during periods of severe market downturns or other crisis situations.
- A wider range of assistance to farmers assessed as not viable in agriculture through short term household support measures.
- Provision for regional and industry adjustment measures that extend beyond on-farm adjustment assistance.
- The establishment of an independent national authority. (derived from Industries Assistance Commission, 1976:79-81)

There are three important points that arise from the IAC's recommendations. First, unlike later versions of the Rural Adjustment Scheme (RAS) (discussed in Chapter Three and Four), the main policy objective for an adjustment program, according to the IAC, was to build up farmers' capacities to improve the allocation of farm resources and to make investments more effective (Industries Assistance Commission, 1976:6). In other words, it was believed that most farmers defined as having a low-income *had the potential* to become more productive in the long term. Here, a welfarist rationality of governing was evident in constituting low-income farmers as members of a broader collectivity who had the capacities to return to a situation of productivity if given sufficient assistance. This contrasts markedly with the rationality underpinning the governing of agricultural change from the late 1980s (see Chapter Three and Four) where only some producers – those with formal managerial skills – were deemed as having the capacities to achieve long-term viability.

Second, the IAC argued that household support, previously unavailable to farmers, should form part of a new rural adjustment program. This sought to recognise that low-income farmers may require a welfare 'safety-net' while deciding whether or not to leave their land. It was assumed here that many low-income farmers would 'know' whether or not they were unviable and take advantage of household support while they made the decision whether or not to leave their properties. This was, again, very different to the late 1980s and

1990s where there was an increased expectancy that unviable farmers would make a 'rational' business decision and move off their land as soon as possible. Third, the IAC recommended that an independent authority be established to coordinate, evaluate and plan rural adjustment at the national level. However, this did not occur until 1992 – some 16 years later – when the Rural Adjustment Scheme Advisory Council (RASAC) was created.

Creation of the Rural Adjustment Scheme (December 1976)

The Commonwealth government accepted many of the IAC's recommendations and established the Rural Adjustment Scheme (RAS) under the *States Grants (Rural Adjustment) Act 1976* (Commonwealth of Australia, 1976a), which came into effect in January 1977. This new scheme brought together all previous rural reconstruction measures into one single integrated program. As noted above, the most significant recommendation rejected by the Federal government was the call for an independent national authority to monitor the administration of the scheme. Both the Federal government and State authorities rejected the option of establishing a national authority as it was believed that it would be more costly with no better results being achieved (Commonwealth of Australia, 1976b). The key aspect of the Review that was accepted was the call for an integrated adjustment scheme. Reflecting in 1996 on their recommendations for the new scheme, the Industry Commission (formerly the IAC) noted that previous reconstruction programs were flawed in not allowing farmers to adjust to demand-based market conditions: 'By not providing opportunities for farmers to adjust to market-determined price incentives, these earlier schemes had not solved the rural adjustment problems' (Industry Commission, 1996:19). The RAS was thus introduced to assist farmers in 'adjusting' to market-determined prices, rather than relying on previous direct forms of support such as output subsidies, tax concessions and fertilizer bounties.

Nevertheless, far from representing a simple shift from government intervention to the abandonment of producers to the vagaries of market forces, the RAS legislation in agreement with the IAC report, acknowledged that a role for government was essential to ensure farmers and graziers had every opportunity to become more productive. According to the Minister for Primary Industry, and National Country Party member, Ian Sinclair, during the second reading on the RAS Bill, the new scheme recognized that government had a responsibility to intervene in improving farm viability:

> ...by providing the financial means to ensure that resources continue to be used in those industries, where their earning power is greatest and, at the same time, provide *welfare assistance* to those farmers so seriously affected by circumstances that they are unable to remain in the industry (Commonwealth of Australia, 1976b:3138, italics V.H.).

Efficiency and welfare reasons were thus argued to justify a role for the Federal government in regulating change. It is worthwhile to note that while the Labor Party, in Opposition, agreed with these objectives, some members did not believe the Rural Adjustment Scheme went far enough in promoting agricultural efficiency. For instance, during the same debate, the Labor Member for Blaxland (New South Wales), Paul Keating (first Treasurer and later Prime Minister 1991-1996), argued that rural adjustment assistance did not go far enough in encouraging restructuring of the rural sector. In many ways, this view was consistent with the rural reforms of the previous Whitlam Labor government. The Labor Party saw assistance to the farming sector as a product of National Country Party patronage and therefore detrimental to national economic efficiency. This was one of the key reasons why the IAC was established. Hence, the Labor Party had an interest in reforms that improved the efficiency of resource distribution in the farm sector.

However, while the RAS could be seen as an economic lever used to dislodge low-income producers from agriculture, another major consideration was to increase the incomes of those producers deemed to be 'viable' (see Lawrence, 1987). There was an underlying assumption, particularly by National Country Party Members that it was the responsibility of government to support farmers in times of low income. In fact, the National Country Party Member for McMillan (Victoria), Barry Simon, noted during the debate of the RAS Bill that it was:

> ...*socially desirable* that farmers be assured of an income when either markets collapse, natural disasters occur, or other matters beyond their control intervene to provide them with a basic income to sustain a lifestyle in rural Australia (Commonwealth of Australia, 1976b:3475, italics V.H.).

The above sentiments were echoed also by Opposition Labor Party members from country seats such as by John FitzPatrick, Member for Darling (New South Wales), who stated:

> Unfortunately, many people seem to think that primary producers are businessmen, the same as any other businessmen; that they make a business decision to go into extra production and if they misjudge the market situation they should suffer the consequences. This argument does carry a lot of validity... [However] they suffer the penalties not only of a bad business decision but also of variations in weather conditions and a lot of other things... [This Bill] will prevent valuable resources being lost and will permit those farmers to remain in the industry, thereby producing further wealth for this country (Commonwealth of Australia, 1976b:3475-3476).

These statements indicate a welfarist rationality of rule in which farmers' decision-making environment is constituted as a risk that is seen to justify a role for authorities in maintaining the collective security of farmers. A welfarist rationality of governing had less emphasis in later versions of the RAS where the viability of producers was constituted as an individual responsibility to be attained through formal managerial skills (see Chapters Three and Four). Thus, in RAS 1977, state agencies were seen still as significant in socializing individual citizenship and economic life according to norms of collective security, albeit for those producers who were deemed to be viable. The objectives of the Rural Adjustment Scheme provide further evidence that a welfarist rationality for those producers defined as viable was maintained. As the Act (Commonwealth of Australia, 1976a:s.26(b)) states:

> The general principle to be applied is to distribute the available resources as widely as practicable, but the over-riding objective is to help restore to economic viability those farms and farmers with the capacity to maintain viability once achieved.

This definition remained the same as that for the 1971 Rural Reconstruction Scheme (see Commonwealth of Australia, 1971a), in which a strong role for state agencies in ensuring as many producers as possible had the capacity to achieve economic viability was also advocated.

According to the RAS legislation (Commonwealth of Australia, 1976a), farmers were to be assisted in adjustment in six main ways. Part A provided assistance in the form of loans at concessional interest rates to long-term viable farmers unable to obtain finance from any normal commercial source for the purposes of:

- Debt Reconstruction – To assist a farmer who, although having prospects of long term financial viability has used all his/her cash or credit resources.
- Farm Build-Up – To supplement, without discouraging, the normal processes under which properties which did not have reasonable prospects of farm viability were amalgamated with an adjoining holding or were subdivided and the subdivided proportions added to adjoining holdings, or to assist a farmer with a property too small to be economic to purchase additional land to build up his/her property to an economic size.
- Farm Improvement – To assist the restoration of an uneconomic property to economic viability by improving the effective use of an existing farm without adding to its area. (derived from Commonwealth of Australia, 1976a)

Also included under Part A assistance in the form of a loan was made available for rehabilitation, to provide limited assistance of up to $5,000 to those obliged to leave the industry where, in the opinion of the administering authority, this was necessary to alleviate hardship. To qualify, the applicant's property must have been purchased by an adjoining owner under the farm build-up provisions or the applicant must have been unable to secure assistance under the debt reconstruction provisions;

Part B provided assistance to long-term viable farmers unable to obtain finance from any normal commercial source for: Carry-On Finance – To provide loans at concessional interest rates for essential carry-on purposes in rural industries which it was agreed between the Commonwealth and a State (or States) were suffering a severe market downturn or similar situation (but excluding circumstances covered by natural disaster arrangements such as drought) (Commonwealth of Australia, 1976a).

Part C provided assistance to non-viable farmers in the form of loans, but convertible into a grant at the administering authority's discretion, for: Household Support – To provide assistance for up to one year to non-viable farmers who had insufficient resources to meet living expenses while the farmer considered whether to leave farming. An extension to two years was possible where a demonstrable effort was made to move out of farming. If the farmer exited the industry the loan could be converted to a grant (Commonwealth of Australia, 1976a). Funding for the RAS was provided by the Commonwealth in the form of loans (85 per cent) and grants (15 per cent). The States were responsible for the Scheme's administration. This excepted carry-on finance, which was funded on a cost-sharing basis between the Commonwealth and the States in such circumstances where this facility was to be used.

The various forms of intervention outlined above show the clear emphasis on Part A assistance, particularly as opposed to Part B or Part C, in terms of the Scheme's key objective of distributing resources as widely as practicable. As noted in a following section of this chapter, the IAC, in its subsequent review of RAS 1977, criticized the excessive focus by both the Commonwealth and State authorities on Part A assistance.

ADDRESSING THE ECONOMIC INEFFICIENCY OF LOW-INCOME FARMS

There was relatively little change between 1977 and 1981 concerning the governing of agricultural change. Commodity prices remained stable and no major new policy initiatives for agriculture were initiated. The principal concern for the conservative Fraser Coalition government toward the end of the 1970s was the effect on agriculture of a forecast economic boom in mineral resources. As agricultural economist Malcolm (1989:139) notes, it was assumed that the appreciation of the Australian currency with the rapid growth in exports of minerals and energy would intensify the economic pressures on farmers – particularly 'adjustment' pressures. The commissioning in September 1981, under a conservative Liberal/National Country Party government, of a Green Paper on agricultural policy represented a response to this assumption.

While the predicted resources boom failed to eventuate, other factors were significant in maintaining the political pressure for a review of agricultural policy. These included, the release in January 1981 by the National Farmers' Federation of a policy document calling for a government White Paper on agriculture (discussed in more detail in a later section of this chapter); revelations of scandal and malpractice in meat processing; and, evidence of inefficiencies in the regulation of the dairy and beef industries (Malcolm, 1989:139). All of these events contributed to an undermining of confidence in the institutional arrangements for agricultural regulation. Released in September 1982, the 'Balderstone Report', named after the Chair of the working group, sought to outline a set of principles for agricultural regulation (Balderstone, Duthie, Jarrett, Eckersley and McColl, 1982). Comprised of public servants, businesspeople and university academics, the Working Group responsible for the Green Paper was significant in constructing an advanced liberal rationality for the governing of agricultural change, much of which was later adopted by the Labor Government, elected in 1983.

The Balderstone Report on Agricultural Policy (September 1982)

The Green Paper argued that it was preferable to specify general aims and objectives for agriculture in the 1980s rather than prescribe growth aims, or the desirable size for particular industries. Consistent with the rationality of other Commonwealth policies, such as the Rural Adjustment Scheme, the Green Paper suggested that agricultural policy should accept the necessity for 'structural' change by:

- being responsive to national and international economic developments;
- creating conditions under which the agricultural sector can handle change and at the same time realise its full potential in contributing to national growth;
- aiming to encourage the efficient use and conservation of the nation's resources;
- promoting the adoption of new technology and increased productivity in the farm sector; and,

- being consistent with other national policies and goals. There should also be consistency of policy at the Commonwealth, State and local levels of government (Balderstone et al., 1982:2).

The report argued that these objectives could be best achieved with minimal direct government intervention. In other words, the Green Paper claimed that the market, rather than governments, would be the most effective means of deciding who were the most productive and efficient producers. However, it also recognised, as had the 1974 Green Paper (Harris et al., 1974), that perfectly operating markets would never be found in practice, and that trade-offs between efficiency and welfare may be desirable in certain circumstances. A welfarist rationality of governing was thereby constructed as important, but used 'too often' as a reason why adjustment should not occur (Balderstone et al., 1982:30).

The Green Paper endorsed some specific economic changes, many of which were recommended in the National Farmers' Federation *Farm Focus* document (for more details see the section on the NFF later in this chapter), and translated into policy by the Labor Government in the mid to late 1980s. These changes included endorsement of low inflation, reduction in industry protection across the economy, and deregulation of the exchange rate, capital market and labour market (Balderstone et al., 1982). Of most significance in the Green Paper, however, was the decisive rejection of tariff compensation as a legitimate form of assistance.

Since the influential 1974 *Green Paper on Rural Policy* (Harris et al., 1974), the tariff compensation argument had dominated debate concerning agricultural assistance. Claiming that primary producers should be assisted in order to offset the reductions in income caused by protection in other industries, it was opposed strongly by the IAC and many economists. However, the Balderstone Report considered that tariff compensation would have significant budgetary costs as well as introduce further distortions in the allocation of resources (Balderstone et al., 1982). In other words, the Green Paper recommended an across-the-board reduction in tariffs. As noted above, tariff reductions were first attempted by the Whitlam Labor government in 1973 and proved to be politically unpopular among farm organizations and the Country Party. With tariff compensation argued by the Balderstone Green Paper to result in further economic and social inequalities, reduced across-the-board government assistance was now constituted as beneficial not only to the efficiency of agricultural industries, but also to improvements in equity of treatment between different industries. The recommendations in the Green Paper involved an approach to regulation weighted heavily in favour of the market as the most effective means of delivering improved efficiency and equity. This perspective, however, was concerned primarily with redistributing resources to those who would use them in the most economically efficient manner. Such a view delimited the possibilities for social forms of intervention and, in the process, assumed that those producers who did not conform to such incentives were inefficient and should leave farming (see Lawrence, 1987).

The Green Paper's emphasis on promoting economic efficiency was subsequently acknowledged by one of the Working Group members, Frank Jarrett, an academic from the University of Adelaide, who wrote:

> I believe that it is an arguable proposition that the failure to make judgments about distributional questions in particular, and economic security questions, perhaps to a lesser

> extent, has in many cases resulted in sweeping some of these questions under the proverbial carpet. I find it very difficult to imagine any policy change, ostensibly made on efficiency grounds, which does not redistribute income in one way or another (Jarrett, 1983:120).

Jarrett here argued that policies encouraging economic efficiency always have an underlying distributional objective, whether this is, or is not, acknowledged. In the case of the Rural Adjustment Scheme this was to encourage, via the farm build-up and farm improvement provisions, that only certain categories of farmers were deemed to be viable. Questions of income distribution in the Balderstone Report tended to be depoliticized by either being categorized as 'welfare' issues – applied only to those farmers about to exit the industry – or constituted as a logical trade-off with efficiency and productivity. As discussed in Chapter Four, such a separation of welfare and efficiency had increasing influence on rural adjustment policy, particularly from the early 1990s. Applied specifically to the Rural Adjustment Scheme, the Green Paper found an ongoing need for this program. In particular, greater emphasis was argued to be required on the farm build-up and farm improvement dimensions of the Scheme. However, it noted that the RAS possibly disadvantaged efficient producers and that stricter monitoring guidelines and control of funding were needed (Balderstone et al., 1982). Farm amalgamation and improved economic use of existing land area were thereby promoted as crucial to a more productive agriculture.

Overall, despite its status as a Green Paper, the Balderstone Report can be seen as a significant document in that many of its recommendations were adopted as policy objectives by the subsequently-elected Labor Government. Given that there were only six months between the Green Paper's release and the March 1983 Federal election, the Liberal/National Country Party Coalition Government barely had time to act on any of its recommendations. Significantly however, the Paper's publication coincided with one of the most severe droughts in the 20th century, which meant that its longer-term orientation was overtaken by more immediate concerns of emergency assistance to farmers (Malcolm, 1989). Once the political imperatives of drought had passed by mid-1983, the Balderstone Report formed an important basis of the Labor government's reformist objectives. The parallels between the Balderstone Report's recommendations, and the political objectives of the Labor Party, are assessed in the following sections.

Rural and Regional Policy: Essay by John Kerin (1983)

A Labor Government led by Bob Hawke was elected with a substantial majority in March 1983. The subsequent reforms pursued by this government need to be seen in terms of the advanced liberal rationality of governing adopted initially by the Labor Party during the Whitlam era, and pursued vigorously by Prime Minister Bob Hawke and Treasurer Paul Keating following the election. As Kelly (1992:20) notes, the Hawke-Keating partnership imposed a new brand of values on the ALP: a belief in market competition, a faith in market forces, a commitment to the internationalization of the Australian economy and reform sanctioned by a consensus process.

These objectives are evident in an article by Minister for Primary Industry, John Kerin, published in 1983 as a series of Labor essays. Written shortly before Labor gained office, Kerin's arguments need to be viewed as a statement on policy (Martin, 1990:160). They are crucial in terms of understanding the rationality of governing underpinning the Labor Party's approach to agricultural regulation. According to Kerin – an agricultural economist by profession – Labor's key objective was to 'create an environment in which efficient exporting industries...will prosper: competitiveness will be enhanced and the well-being of producers increased' (Kerin, 1983:65). Clearly, the decisions in 1983/84 to float the Australian dollar and deregulate the financial system, and later to allow the entry of foreign banks, reflected one of the outcomes of this desire to establish a competitive environment. These decisions followed the Martin and Campbell Inquiries into the finance system and assumed that the well-being of producers would not conflict with and, in fact, was compatible with improved competitiveness – a key assumption of subsequent attempts by the Labor leadership to pursue an economic reform program (see Chapter Three). There was also an enhanced emphasis on economic efficiency, and greater 'responsiveness' of policy to changing international conditions. This was a central recommendation of the Balderstone Report.

For Kerin, Labor's approach to the review of existing and proposed agricultural assistance measures was to be based on three criteria:

- Does the assistance measure contribute to the flexibility of agricultural industries, or is it likely to hinder the adjustment process by insulating individuals from the economic and social realities of the environment in which they operate?
- Does the amount spent on individual assistance measures represent the best 'value for money' for primary producers, or could the same amount be spent in a different way and make a different contribution to the prosperity of agriculture?
- What are the distributional aspects of the assistance measure? The prime consideration in determining the distributional impact of any assistance measure is not equity in an absolute sense, but whether the impact of the assistance is consistent with its objectives. (derived from Kerin, 1983:60-61)

These objectives have a strong focus on more 'active' forms of conduct on the part of producers in order to adjust to claimed economic realities, and a rejection of social equity as an end in itself. A welfarist rationality of governing was, in this essay, constituted as a hindrance to change and argued to shield farmers from market realities. This calling into question of a welfarist form of governing was again emphasized in 1984 when the IAC released its review of the Rural Adjustment Scheme.

IAC Report on Rural Adjustment (1984)

The IAC was commissioned in late 1983 to review the operations of the Rural Adjustment Scheme and to report on whether any change should be made to, or additional measures introduced to meet the longer term adjustment needs of, individual rural industries (IAC, 1984:1). In April 1983 the IAC had released an Interim Report on the immediate needs of rural industries resulting from the drought then affecting much of eastern Australia. This report recommended that any changes made to the Rural Adjustment Scheme would not improve efficiency of resource use during a period of drought (IAC, 1983), a recommendation subsequently accepted by the Federal Government. However, the IAC's

1984 report on longer-term adjustment argued for a number of changes to the operation of rural adjustment policy.

The IAC found a number of problems with the RAS. These problems are outlined below.

1. Most assistance had been provided under Part A (debt reconstruction, farm build-up and farm improvement) with very little uptake of rehabilitation and household support.
2. A significant proportion of the concessional element of Commonwealth funds had been retained by the States.
3. There had been inadequate monitoring of applicants' financial performance before or after receipt of RAS assistance, with little information on what farmers assessed as 'non-viable' or 'not in need' had actually done with their resources.
4. It was difficult to identify whether those farmers for whom the scheme was designed had received benefit. Evidence from the Bureau of Agricultural Economics (BAE) suggested that the majority of assisted farmers would have been able to raise funds from normal commercial sources.
5. The subjectivity required in assessing eligibility for RAS assistance increased the difficulty of defining the characteristics of those farmers expected to benefit from the scheme.
6. RAS had operated largely as an adjunct to the commercial finance market and had failed to improve its efficiency by relieving commercial lenders of their responsibilities for providing credit to farmers.
7. Some of the assistance provided through concessional credit was likely to have been capitalized into land values.
8. The gains in efficiency and welfare of providing selective adjustment assistance to some farmers were seemingly wasteful for the community as a whole.
9. There was no evidence that RAS assistance filled gaps in the social security system by providing general welfare support to low income farmers. (derived from IAC, 1984)

These observations on the Scheme's operation led the IAC to argue that the RAS was not fulfilling the Federal government's stated objectives. As with their 1976 review of rural reconstruction, the IAC's findings were mediated by their legislative objectives as a statutory body. Thus, the Commission stated clearly that any assistance policy, such as the Rural Adjustment Scheme, that dealt with the viability of particular industries 'should improve the efficiency with which the community's productive resources are used, and to facilitate adjustment to changes in the economic environment by persons and industries affected by those changes' (IAC, 1984:42). In other words, in evaluating the RAS, the IAC's primary concern was the efficiency of farms as an element of broader national economic management.

Nevertheless, this was not entirely uncontested. For instance, the New South Wales Department of Agriculture – then working under a Labor government led by Neville Wran – argued that the RAS should be evaluated according to its capacity to compensate farmers disadvantaged by change:

> The adjustment problem becomes less a matter of efficiency with which the nation's resources are used and more a matter of welfare based on equitable concepts of income

> distribution and compensating people who are disadvantaged by the adjustment process (cited in IAC, 1984:47).

This statement appears to continue to favor a welfarist rationality of governing which was, by the early 1980s, subject to widespread questioning in Australian politics. The promotion of economic efficiency, endorsed by the IAC – as a form of expertise on economic matters – was constituted increasingly as central to the task of governing.

The IAC took into account two different views in considering whether the RAS should be continued. The first approach saw the scheme as another distortionary instrument which could be used in response to the demands from pressure groups, while the second view supported the RAS on the grounds that it removed pressure for more distortionary and costly ad hoc assistance measures (IAC, 1984:70). The second view was argued strongly by the Federal Department of Primary Industry, the New South Wales Department of Agriculture and the National Farmers' Federation. However, the IAC adopted a position that sought to reconcile these two competing perspectives. Overall, it found that the objective of RAS 1977 was 'consistent with broader social policy rather than of economic efficiency' (IAC, 1984:47) and, in fact, had the effect of encouraging the provision of additional resources to those farmers who might not otherwise have a long-term future in the industry.

On this basis it recommended:

- That the Rural Adjustment Scheme be discontinued as a generally available scheme for farmers meeting the eligibility criteria, but be available for farmers in a particular region or industry only after an independent public inquiry into the effect of such assistance on the development of an improved industry structure.
- That assistance be provided for a specified time only, and then be terminated or publicly reviewed.
- The scheme provide for farm-level adjustment assistance only through the existing measures of the RAS.
- That assistance should be provided in the form of grants or interest subsidies rather than concessional loans.
- That the eligibility criteria of the existing scheme be retained.
- That adjustment assistance programmes be monitored by an independent national body. (derived from IAC, 1984:76).

There were three important changes suggested by the IAC to the Rural Adjustment Scheme. First, and perhaps most significantly, the IAC recommended that the RAS be discontinued as a generally available scheme, but should be retained in certain circumstances. This half-hearted support of the RAS reflects an increased focus by the IAC on economic efficiency. In particular, the IAC seemed to have been strongly influenced by the outcomes of the Balderstone Report in which greater conformity by farmers to market signals was favored. Thus, while the RAS was to be retained in a minimalist form, it was argued that any assistance provided should be conditional on improvements in industry economic structure. This represents a shift from RAS 1977 where both the efficiency of farms and welfare of farmers were considered crucial in the governing of low-income farms. There was now a much more explicit focus on ensuring that economic efficiency had primacy in considering whether assistance should be made available.

Second, the IAC expressed a preference for assistance in the form of grants and interest subsidies rather than concessional loans. The justification for this was that concessional loans had resulted in State authorities committing most of their resources to providing lending facilities with only a small proportion of the monies actually reaching farmers (IAC, 1984:73). There was also evidence that many States had been charging farmers higher interest rates on loans than those specified by the Commonwealth. This issue was raised subsequently during parliamentary debate by Federal Labor Member for Capricornia (Queensland), Keith Wright, who argued that the conservative National Party government in Queensland had been 'robbing rural producers blind' through concessional loans (Commonwealth of Australia, 1985b:2749). On this basis, the IAC suggested that direct grants and interest subsidies provided the greatest flexibility to farmers, were less expensive for State authorities to administer, and allowed for greater transparency and accountability. It was assumed by the IAC therefore that such transparency of process would make the RAS operate more effectively by allowing for improved monitoring by the Commonwealth. However, a consequence of such flexibility was an increased onus on both State authorities and farmers to account for the 'efficient' use of such funding. This, in fact, became an important focus of the 1988 version of the RAS (see Chapter Three). Third, as in 1976, the IAC again recommended the creation of an independent national authority for monitoring the RAS. As with the IAC's key recommendation of discontinuing the RAS, its argument for a national authority was again rejected by the Federal Government.

Revised Rural Adjustment Scheme (June 1985)

A revised Rural Adjustment Scheme was assented to on the 6th of June 1985 operating under the *States and Northern Territory Grants (Rural Adjustment) Act* (Commonwealth of Australia, 1985a). As noted above, the Federal Government rejected the IAC's recommendations that the RAS be discontinued as a generally available scheme and that an independent national monitoring body be established. The government recognized that the RAS should remain in a similar form to 1977 but that some changes were necessary to funding arrangements.

These changes, which were consistent with IAC proposals, included the following:

- Under Parts A of the scheme, the form of adjustment assistance was changed from concessional loans to interest rate subsidies of up to 50 per cent. Such funding was made available to the States and Northern Territory wholly in the form of grants.
- Under Part B of the scheme, the form of carry-on assistance was changed to interest rate subsidies of up to 50 per cent with the Commonwealth and States/NT each contributing 25 per cent.
- Rehabilitation assistance was moved from Part A to Part C of the scheme with the maximum amount of assistance increased to $8,000 and linked to the Consumer Price Index (CPI). The assistance remained as a loan convertible to a grant at the discretion of State authorities. (derived from Commonwealth of Australia, 1985a)

The changes in funding were argued by the Minister for Primary Industry, John Kerin, to be important in introducing greater transparency and flexibility in administration of the RAS. As the Minister stated during the Parliamentary debate on the IAC Bill, 'the assistance provided by the Government, in the form of an interest rate subsidy for eligible farmers, will be much more visible and the benefits will be passed on to farmers' (Commonwealth of Australia, 1985b:1868). However, in the same debate, National Party Member for Dawson (Queensland), Raymond Braithwaite, advised caution in this approach and argued that more realistic guidelines needed to be adopted by the State adjustment authorities so that the needs of farmers were properly addressed (Commonwealth of Australia, 1985b:2753). The issue of State authorities conforming to Commonwealth Government objectives will be examined more closely in the next section of this chapter.

In line with the changed funding arrangements there was also an enhanced emphasis on commercial viability in the Rural Adjustment Scheme's objectives. This reflected the recommendations of both the Balderstone Report and the IAC Review of the RAS. However, a concern with distributional issues was maintained in the final legislation:

The general principal to be applied is to distribute the available resources as widely as practicable... The overriding objective is to assist rural industry structural adjustment and to ease adjustment pressures through the provision of adjustment assistance to individual farmers who are considered to have the capacity to achieve and maintain a commercially viable farm business enterprize (Commonwealth of Australia, 1985a:Schedule, Part 1).

While recognizing that as many farmers as possible should have the opportunity to remain in the industry – providing that they had the capacities to become commercially viable – an enhanced emphasis was placed on directing assistance to those farmers who were assessed to constitute their farms as business enterprizes. The building of farmers' *business capacities* was here constituted as an important part of achieving viability; a point developed in Chapter Three. This is a crucial shift in governmental rationalities from the RAS 1977 objectives, and represents one of the first clearly articulated attempts to govern farmers in an advanced liberal way.

'Broad' Interpretations: Economic Management and the State Administration of Farm Adjustment

As noted in Chapter One, State administrative authorities have played a fundamental role in the assessment of applicants under the Rural Adjustment Scheme, and thus in governing agricultural change. While the specific assessment procedures of State authorities are not examined in this book, it is useful to consider how Commonwealth agencies have sought to govern the conduct of State authorities to achieve conformity with national objectives.

Relations among Commonwealth and State reconstruction/adjustment agencies between 1973 and 1984 have followed a general pattern of flexibility (in practice), with judgment on farm assessments a matter for State administrations – defined through the guidelines in the Federal-State agreement. However, from the outset this flexibility has been deemed a problem in achieving desired results. The problem of administering a national policy in six different jurisdictions (the Northern Territory was not included until 1985) had been noted by the Industries Assistance Commission in its 1976 review of the Rural Reconstruction Scheme, when it suggested that there had been 'a general lack of coordination and liaison between the Commonwealth and the States leading to different treatment of individuals between States'

(IAC, 1976:4). The IAC argued that despite equity levels, debt structure, level of risk and uncertainty, and the ability of the operator all forming a part of how State authorities assessed applicants, these factors varied in their type and emphasis.

While the IAC argued that it was difficult to determine the degree to which differing assessment procedures had resulted in a lack of uniformity, variation was suggested to arise from:

- Differences in the background and nature of the administering bodies.
- Differing interpretations of the objectives and provisions of the legislation.
- Differences in the farming conditions and environment in which the State authorities have operated. (derived from IAC, 1976:34-35)

The suggested solution to this problem was a more 'objective' assessment of viability through farm inspections, thus providing a claimed accurate picture of the present situation and future prospects of the farmer. As viability was formally defined in neither the 1977 nor 1985 versions of the RAS, it was very much a matter of practical (and subjective) judgment by the State authorities based on the ability of applicants to service their debt repayment commitments. Through an emphasis on farm inspections as part of the assessment process, and improved farm management and technical advice, the IAC believed that the State authorities could maintain their responsibility for judging the future prospects of applicants, while taking into account the agreed principles and guidelines of the Federal-State agreement (IAC, 1976:68). However, farm inspections were not incorporated into the final RAS 1977 scheme.

Perhaps not surprisingly, the IAC's report on rural adjustment in 1984 found that the determination of 'viability' was still very much a contested matter. In fact, it found that there existed 'considerable disagreement' as to whom the scheme was intended to assist (IAC, 1984:49). While the Australian Bureau of Agricultural Economics (BAE), during hearings, argued that objective parameters for determining viability could be determined using a farm's equity-income ratio, this was disputed by RAS authorities. The State-based RAS administrative bodies tended to take a more 'generous' view of viability than the BAE, IAC, or Commonwealth government concerning the types of farms and farmers in need of assistance (IAC, 1984:49).

Such broad interpretations of the RAS provisions by State authorities presented a problem from a Commonwealth perspective. For example, as John Kerin, Minister for Primary Industry, in the Parliamentary debate on the States and Northern Territory Grants (Rural Adjustment) Bill noted:

> I will not forget the first time when, as Minister for Primary Industry, I attended a meeting of the States on the rural adjustment scheme. I asked the States to look more seriously at household support and to interpret that a little more liberally because it was my estimation that some industries would be in serious trouble. The States were not inclined to look at the matter at all. Basically, they believed that the rural adjustment scheme was for more traditional forms of keeping farms viable. They would not look at other elements of the scheme (Commonwealth of Australia, 1985b:2754).

The above statement reflects the frustration of the Minister in getting State authorities to interpret 'correctly' Commonwealth objectives. It appears from the Minister's comments that the States had quite different ideas from the Commonwealth concerning how RAS monies should best be targeted, and which farmers required assistance. The State authorities seemed to have a concern with ensuring as many farmers as possible were categorized as viable, thereby focusing their efforts on the farm build-up and farm improvement aspects of the RAS. Thus, while a welfarist rationality of governing was open to question by Commonwealth agencies, many State authorities, through their assessment procedures, sought to govern farmers in a welfarist way.

In interpreting the RAS guidelines more liberally than the Commonwealth might have liked, the State authorities were effectively asserting the political complexities of defining, and thereby governing, viability. This affected the operation of the entire Scheme from the outset which, in turn, led to calls for more effective strategies to monitor the assessments of State authorities; hence the introduction of interest rate subsidies in 1985 to replace concessional loans. The strategy of using different technologies of financial management for the Commonwealth to more effectively govern the conduct of State authorities 'at a distance' was by no means new. Indeed, the Balderstone Report had recommended that 'effective and coordinated action will often require that the Commonwealth provides financial incentives to encourage independent State governments to participate and cooperate (Balderstone et al., 1982:8). As the Parliamentary statement by John Kerin made clear (see Commonwealth of Australia, 1985b), the introduction of interest subsidies was aimed towards achieving greater visibility of process by State authorities, and thus improved monitoring by the Commonwealth. This occurred again in 1988 when the provisions for financial monitoring and accountability were further re-evaluated as part of attempts by the Commonwealth to achieve improved 'accountability' by State authorities.

The Management of RAS Assistance in Queensland

Throughout the period from 1973 to 1985 the administration of the Rural Adjustment Scheme in Queensland was the responsibility of the State Rural Reconstruction Board (RRB). During this time a conservative National Country Party was in government. The RRB was created originally in 1970. It comprised public servants whose task was to administer assistance provided through the Commonwealth Rural Reconstruction Scheme. This later extended to administration of the RAS. However, by 1985 the National Party regarded the RRB as no longer appropriate to the operation of rural industries in Queensland, and its role became incorporated under the new Queensland Industry Development Corporation (QIDC).

The QIDC was created in 1985 following two events. First, a task force on employment in Queensland released a report in early 1985 recommending that assistance in expanding the State's economic base was imperative in the facilitation of economic activity. Second, a policy speech made in 1983 by the Deputy Premier, William Gunn, argued that additional assistance would be provided 'to the rural industry, small business and pioneer industries' (cited in Parliament of Queensland, 1985:3456). On the basis of these two events, the governing National Party introduced a Bill to establish an industry corporation – the QIDC. The main objective of the QIDC was:

> ...to facilitate, encourage and promote the development and expansion of primary, secondary and tertiary industries in Queensland with a view to enhancing economic

growth and employment opportunities in the State (Parliament of Queensland, 1985:3456).

The creation of the QIDC represented a major challenge in Queensland to the almost fundamentalist pre-occupation with assisting primary industry (see Scott, Coaldrake, Head and Reynolds, 1986). It marked a shift toward a rationality of governing where businesses displaying 'the best prospects for growth' (Parliament of Queensland, 1985:3457) were encouraged. Such an objective was consistent with the objective of the RAS. This was reflected in the replacement of the Rural Reconstruction Board by a Rural Adjustment Advisory Board, which was established as a subsidiary body of the QIDC.

While the purpose of this Board was not fundamentally different to the former RRB, its composition of all business people, and its control by the QIDC Board, meant a much stronger private sector orientation. It is difficult to speculate on the effect that this had on the assessment of RAS applicants. In fact, evidence from Queensland public servants and subsequent RAS reviews suggests that there was little change. However, the Queensland government's attempts to expand industry incorporated a concern with 'rational' economic management – based on a corporatized structure for administering assistance – hence, seeking to align the administration of assistance more closely with Commonwealth objectives. A shift toward improved economic management was evident also in attempts during the late 1970s to form a single political voice for farmers.

Farm Organizations and the Assemblage of a National Voice

The years 1973 to 1985 represent a period in which a national voice for primary producers emerged through the National Farmers' Federation (NFF), formed in 1979. An economically 'dry' organization, the NFF is dominated by the free trade interests of grazier groups – particularly large woolgrowers (see Lawrence, 1987; Connors, 1996). It is the only national body 'responsible for developing and implementing policy on matters affecting the rural sector as a whole and on matters affecting more than one commodity' (cited in Trebeck, 1990:138). The NFF consists of a national executive, and representatives from State-based farming organizations and commodity councils. This apparent unity has given the NFF considerable political influence on matters of agricultural policy (Lawrence, 1987; Martin, 1990). In large part, unity has been maintained due to the central policy principles of the NFF executive – freer trade, flexible wages, financial deregulation, reduced protection, a focus on competitiveness and reduced government expenditure – endorsed as being in the interests of all agricultural producers.

However, this has tended to marginalize the pro-interventionist concerns of small commodity producers (Lawrence, 1987:217) represented traditionally by the National/Country Party. Rather than the National Party seeking to protect this traditional constituency, it too has moved progressively to the Right to maintain its political relevance, comparative to the NFF, in issues concerning agriculture. The rise to prominence of the NFF has been covered in detail by other authors (see specifically Connors, 1996). It is not my intention here to repeat this material. Instead, I focus on some of the events and forms of knowledge that enabled the NFF to be assembled into a coherent form for speaking authoritatively on agricultural change.

Perhaps the most important event that enabled the NFF to emerge as a politically potent organization were the increasing levels of professionalism by farm organizations in the early 1970s. At this time there was already a shift from the idea of 'protection all round' to the evaluation of assistance measures through the use of formal economic techniques (see Martin, 1990). This trend was enhanced considerably when, in 1974, the IAC was created. The IAC was important in both defining the issues that farmers faced, as well as constituting how these issues should best be resolved (see Trebeck, 1990). From the perspective of an analytics of government, the IAC represented a centre of calculation in constituting various issues as problems, rendering these problems into a governable form, and suggesting 'appropriate' responses by producers. This led to two significant developments.

First, it had the effect of promoting greater unity between what was up until this time a wide diversity of different commodity groups (Trebeck, 1990; Connors, 1996). The creation of the IAC added additional weight to the arguments of those who sought unity, particularly in terms of representing producers interests according to a 'core' set of economic objectives (see Trebeck, 1990). Second, it strengthened the move toward the hiring of professional economic experts by farm organizations to assist in the research and writing of written submissions (Marshall, 1986). The need for submissions to incorporate 'sound' economic arguments led many farmer organizations to hire full-time trained economists, rather than part-time lobbyists. This served to meet the criteria for acceptable submissions and also to frame submissions increasingly in terms of 'rational' economic management.

The use of economic arguments was suggested by the Federal government, IAC and economists to represent a more accurate and truthful means of assessing agricultural assistance levels than the political patronage of the past encouraged by the National Country Party. Of course, this did not stop farm organization representatives going to Ministers, but this now was seen as an ineffectual way of influencing policy, as the Ministers too were operating under the formal guidelines. Perhaps not surprisingly, many farmer organizations found it difficult to adjust to the new circumstances. Marshall (1986:30) notes that a number of groups initially found the task burdensome and called on the Federal government for assistance in making claims for assistance! This changed over time, and today all farm organizations in order to be politically effective have economic expertise that specialize in the making of submissions. Nevertheless, at the time, formal guidelines enabled the emergence of not only a shared expert discourse among farm organizations focusing on issues of economic efficiency, but also an increased capacity for alignments in rationalities and technologies – via the requirement of 'rigorous' economic arguments – between these organizations, economists and formal political actors.

The increasing formalization of policy submissions accords with a shift toward governing agricultural change in an advanced liberal way. Certainly, the formalization of submissions created an environment in which agricultural economists could have improved access to government. It served also to align the vocabularies, material resources and forms of calculation among farm organizations so that they constituted problems in a similar way, and represented issues in an ordered form. This was important in increasing the pressure for unity. Once organizations used similar discourses and modes of calculation that accorded with government-based goals of prudent economic management, their capacities to act in the formal political sphere were enhanced considerably (see Murdoch, 1995 for a similar argument in relation to UK agriculture). Conversely, formal unity between different interests enabled government agencies to monitor the activities of farm organizations 'at a distance' (Miller and Rose, 1990) through the expert-defined calculative techniques of agricultural economists.

While the creation of the IAC enabled economic experts to assume a more prominent role within farm organizations, this was not of itself sufficient to establish a single body. The quest for unity was driven primarily by grazier organizations in New South Wales who sought to avert the squeeze on agricultural incomes through lower costs, and improve profits through the removal of claimed distortions in resource allocation between sectors (Trebeck, 1990:142). Organizations such as the United Farmers' and Woolgrowers' Association and the Graziers' Association of New South Wales – who amalgamated in January 1978 (Trebeck, 1990:136) – argued that a strong voice at a Federal and State level would enable these goals to be pursued while, at the same time, improving the financial resources of farm organizations (Connors, 1996:197). Unity was thereby suggested to promote, at a national level, a pooling of resources to encourage the key issues of reform in agricultural protection and more 'flexible' labour policies.

The live sheep dispute in 1978 was important in demonstrating the potential benefits of unity. This dispute brought together a number of producer organizations against a common enemy – the Australian Meat Industries Employment Union (AMIEU). Briefly, the live sheep dispute erupted between meat workers and livestock producers over the shipment of live sheep to the Middle East. While the export of live sheep raised the price of mutton for producers, it effectively threatened employment in the meat processing industry. In response to the exports, the AMIEU established picket lines around holding pens at various Australian ports preventing the sheep from being loaded. However, through tactics such as transferring sheep to rural ports for loading, a public relations campaign to gain public support, and pressure on the Federal and State governments, livestock producers eventually defeated the unions (Connors, 1996:205-206). The gains achieved collectively in this dispute demonstrated the potential political strength of a national representative body. In particular, it showed the types of issues that the large broadacre organizations leading the push for unity were interested in pursuing. A decade later this had significant implications for the enthusiasm of the NFF to 'take on' trade unions (Malcolm, 1989). Following the live sheep dispute, although not caused by it, amalgamations occurred between farming and grazing organizations in Victoria and South Australia.

However, the Cattlemen's Union in Queensland opposed the shift during 1978 and 1979 to Federal unity on the grounds that the agenda was dominated by wealthy woolgrowers – who wanted an across-the-board reduction in government assistance.[2] The eleventh hour agreement by the Cattlemen's Union (CUA) was significant therefore in enabling the NFF to speak on behalf of cattle producers. The Cattlemen's Union was formed in 1976 following a bitter dispute with the United Grazier's Association (UGA) over meat marketing arrangements during the slump in beef prices between 1974 and 1978 (Schmalkuche, 1986). From its beginnings the CUA was regarded as radical, seeking assistance from government in an otherwise unprotected and free enterprize industry.

The CUA did not get involved in the live sheep dispute, but signed a deal with the AMIEU that restricted live cattle exports to those of its members (Connors, 1996:206). This

[2] Lawrence (1987) provides a critical evaluation of the agenda underpinning the push for unity.

caused a great deal of animosity towards the CUA. The CUA's ideological opponents were the wealthy graziers (represented by the UGA), who opposed meat marketing reform and endorsed minimal government intervention. It was the claim by the CUA president that the unity movement was dominated by the sectional interests of graziers intent on pursuing a free trade agenda that initially prevented CUA affiliation with the NFF (Connors, 1996:209). Even if the free enterprize position of the woolgrowers did later prevail in the NFF, the CUA in 1979 had to decide whether to join, and have an active role in the policy process, or not join and become largely irrelevant – given that many of the other main producer organizations had already entered the fold of the NFF. It opted for the former and, through an arbitrator, was able to gain three out of the six Queensland Cattle Council seats on the NFF (Schmalkuche, 1986:120).

This was a significant move as it meant that the NFF now had no competition in claiming to represent the interests of cattle producers. Once the CUA was part of the NFF, any calls for protection could more easily be monitored and de-legitimized as not reflective of the organization as a whole. In other words, the NFF could establish itself as an effective centre of calculation for 'farm interests'. Thus, Martin (1990:158) notes that while the NFF have tended to support economically rational policies, its constituent commodity councils – of which State-based organizations are members – favor 'distortionary' policies that support their particular interests. In the two decades since its creation, the NFF executive's commitment to economic reform seems to have prevailed at the expense of the voices in its affiliated organizations (see Marshall, 1986). It has moderated the concerns of its affiliated commodity councils by compacting diverse interests into a singular rationality of governing based upon competitiveness and freer trade.

The NFF could thereby claim to speak 'with one voice' (Connors, 1996) by the end of 1979, thus enabling it to act as a centre of calculation, or an 'obligatory point of passage' (Callon, 1986), between farmers and state agencies. In fact, the emergence of the NFF as the exclusive voice of Australian farmers shows many parallels with the development during the 1940s in the United Kingdom of the National Farmers' Union (NFU) (see Murdoch, 1995). Like the NFU, the NFF's incorporation into the policy process was enabled through the construction of farmers' interests in a manner that unified competing and often contradictory demands. In this way, the NFF acted as a key agent of governmentality, condensing diverse interests and representing them as those of the entire industry. In 1981 the NFF's political aspirations were made clear when it released a statement of objectives on farm policy. As Trebeck (1990:139) argues, this document was significant in cementing the NFF's position not only in agriculture, but also in the wider debate on economic regulation.

National Farmers' Federation – *Farm Focus: The 80s*

The NFF released *Farm Focus: The 80s* at a time when more formal economic approaches to the analysis of policy options were being sought by government (Marshall, 1986; Malcolm, 1989; Martin, 1990). This paper argued that ad hoc policy responses in the rural sector needed to be replaced by long-term responses that recognised an expanding role for agriculture in the Australian economy. The NFF, as part of its goal to institute a more rational approach to addressing agricultural problems, called for the preparation of a White Paper on agriculture. While a White Paper never eventuated, the Federal government agreed to commission a working group on agriculture (Trebeck, 1990:139). The Balderstone Report (Balderstone et al., 1982) represented the outcome of this working group's deliberations.

In order to achieve a long-term agricultural policy 'consistent with the interests of the entire economy', the NFF suggested the following policy objectives.

- Increase farm profitability;
- Improve the competitiveness of farm products;
- Achieve a major expansion of Australian agriculture;
- Improve productivity and adjustment;
- Achieve equity for rural communities; and,
- Encourage conservation of the environment (derived from NFF, 1981:2-3).

If considered together, the objectives of profitability and competitiveness seem contradictory to those of environmental protection and equity in rural communities. Nevertheless, the NFF believed that these objectives were compatible and could be achieved concurrently. Farm adjustment represented one of the central strategies through which the NFF argued that these objectives could best be pursued. The purpose of adjustment, according to the NFF, was 'to 'oil the wheels of change'; to assist, rather than impede, market forces' (NFF, 1981:53). In other words, previous government policies were constructed as impeding prudent economic management and inhibiting the exit of those farmers defined as not having long-term prospects. This position is not unlike that endorsed in the Balderstone Report. As the NFF (1981:53) notes, 'the role of government should be to facilitate adjustment, rather than planning and directing it, or worse, preventing it from occurring'. What this statement failed to recognise was that by attempting to define, and govern, adjustment, governments (and the NFF), whether they liked it or not, were directing the form and pace of change.

A further significant aspect of the NFF paper was its emphasis on social and non-economic goals as being conditional on the achievement of economic objectives. This was emphasized by various authorities from the late-1980s (see Chapters Three and Four). However, from the late 1970s until the early 1980s the conservative Liberal/National Party Coalition government remained wedded to a welfarist rationality of governing; at least for farmers. While the Federal government encouraged 'prudent' economic management, recognition of the social value of farming remained important politically in maintaining the National Party's electoral support among small commodity producers. Thus, blanket endorsement of the NFF's position would have proved political suicide for the National Party, and threatened its place in the Coalition. Election of a Labor government in 1983, and the National Party's move to the political Right enabled the NFF's objectives to be more successfully pursued in the late 1980s.

The political rationality informing the NFF's recommendations on economic efficiency was embraced strongly by the Hawke Labor government following their election in 1983. As Trebeck (1990:139) argues, the debate over protection for agricultural industries was largely won during the 1980s due to the Labor government's acceptance of deregulation and international competitiveness. For instance, the NFF's call for deregulation of the credit market, as a means of improving adjustment, was not achieved until after the Hawke Labor Government was elected. Overall, the NFF's recommendations were closely aligned with the Labor government's farm objectives throughout the 1980s and, in fact, were adopted widely following the 1986 terms-of-trade collapse when previous forms of governing the economy were argued to be incapable of enhancing international competitveness (see Chapter Three). Arguably, none of this would have been possible if the NFF had been unable to act as a centre of calculation, representing a heterogenous network of commodity councils and State-based farm organizations with one voice.

CONCLUSION

From 1973, low-income farms were constituted increasingly as a widespread and long-term problem of economic efficiency. This emerged largely through the creation of the IAC, which encouraged two trends. First, it introduced formalized technologies of calculation and material representation as a more 'rational' means of evaluating policies of assistance. These technologies included particularly formal submissions to argue for the costs and benefits of particular forms of intervention. Second, these technologies enabled agricultural policies aimed toward regulating agricultural change to be evaluated in terms of their effect on the economy as a whole. Third, the creation of the IAC accorded economists authority as experts in constituting farm problems, and in transforming these problems into a discourse of economic efficiency. This questioned the effectiveness of welfarist rationalities of governing that had sought to preserve the economic welfare of specific industries via collectivized forms of assistance. In fact, the processes whereby particular industries had received assistance were now open to intense scrutiny, and regarded as a serious problem in achieving optimal functioning of a national economy. Following its formation in 1979, the National Farmers' Federation sought also to question previous ways of governing the conduct of farmers. The NFF drew extensively on economic expertise and argued strongly for governments to ensure that farmers adjusted to claimed market realities in the interests of freer trade and a more competitive sector.

This was not, however, entirely uncontested. The National Country Party, during its term in office with the Liberal Party (1975-1983), sought to maintain welfarist forms of assistance (at least to those farmers deemed as viable) to keep the support of its traditional constituency of family farmers. An uneasy tension thus existed between objectives to improve the economic efficiency of farmers, and the political desire to ensure that assistance measures to farmers were maintained. This rationality of governing incorporated elements of both welfarist and advanced liberal forms of rule. Following the election in 1983 of the Hawke Labor government the social rationality for maintaining farmers on their land was deemed less significant. The Hawke government sought to pursue the reforms that had been commenced during the previous Labor period of rule (1972-1975), and focused more intensely on improving economic efficiency by building the managerial capacities of farmers.

The next chapter examines the period from 1986 to 1989 in Australian politics and the rationalities and technologies of governing through which the managerial practices of farmers were able to emerge as the proper sites of reform. Of particular significance is how the construction of an economic crisis, and attempts to internationalize the economy, enabled farmers' individual attitudes and capacities to emerge as central elements in securing national agricultural efficiency and progress.

Chapter 3

Creation of the Enterprizing Farmer: International Competitiveness through 'Outward-Looking' Attitudes

The previous chapter outlined how economic expertise assumed significance from the early-1970s in speaking 'truthfully' on the issues of change affecting Australian agriculture and, specifically, those farmers seen to have little future in the industry. It emphasized particularly the need to provide forms of assistance that enhanced the efficiency of the entire economy as a whole rather than the welfare of particular industries. Such a concern was extended during the late-1980s as attempts were made to internationalize the Australian economy and to improve the competitive position of agricultural industries. However, rather than enhance competition through older styles of industry support, the type of reform beginning to emerge emphasised farmers attitudes as the legitimate and most effective site through which economic efficiency could be pursued. This shift had considerable consequences for the individualizing of farm management problems and the types of farmers now seen as 'viable'. Commencing in 1986 with Treasurer, Paul Keating's, notorious 'Banana Republic' statement, the chapter traces the emergence of this 'outward looking' approach to agricultural restructuring, which sought to govern agricultural change through the individual capabilities of farmers.

Farmers' attitudes assumed prominence as a problem from the 1980s through a series of statements calling into question previous forms of governing economic and agricultural matters. This problematization of rule emerged specifically during late 1985 and 1986 when national terms of trade sank to historically low levels (see Kelly, 1992; Gruen and Grattan, 1993). On the basis of a claimed economic crisis, previous forms of (Keynesian) economic intervention were called into question and blamed for inhibiting growth, inducing farmer dependency on government subsidies and preventing international competition. A representation of crisis in the economy, and attempts to govern it, enabled the economic capacities of farmers to be constructed as a site of reform and restructuring. Rationalities of governing shifted from a concern with questions of 'How can government intervene in economic life to create desirable social outcomes?' to 'How can government create competitive market situations to ensure world-best practice is pursued and international standards are achieved?' (Catley, 1996:65).

A shift in rationalities of governing was reflected in the encouragement of a business-focused approach by farmers as a potential solution to improving farm productivity and viability, and thus the international competitiveness of Australian producers. Also of interest during this period was an attempt by the Commonwealth government to introduce greater administrative 'transparency' by State authorities in their administration of the Rural Adjustment Scheme (RAS). Both of these changes were crucial in enabling greater economic surveillance over not only the conduct of farmers but also the procedures of administrative authorities.

(DIS)ORDERING THE ECONOMY

In Australia, the years 1986 to 1989 represent a period of significant change in managing not only agricultural change, but also the economy more generally. Such changes reconstituted the image of the Australian economy from one seen as national, self-contained and ordered, to one that was disordered, crisis-ridden and integrated in a competitive international order (Beeson and Firth, 1998; Hindess, 1998). This sense of national economic crisis was important in calling into question the capacities of state agencies to govern effectively, and in focusing attention on shaping the managerial capacities of farmers in order to resolve this crisis. Two key statements during 1986, the White Paper on *Economic and Rural Policy*, and the Federal Treasurer's Banana Republic statement, were crucial in constructing a discursive framework of reform through which both the activity of governing, and the capacities of individuals and industries, could be constituted and shaped in new ways.

Statement on Economic and Rural Policy (April 1986)

The White Paper on *Economic and Rural Policy* was released at a time of increasing costs (mainly through high interest rates) and low returns for Australian farmers. It represented a comprehensive review of economic and rural policy in order to evaluate how these financial problems might be addressed. The broad strategies put forward as a way of dealing with farm sector problems included:

- The achievement of sustained economic growth through lower inflation.
- The development of a more outward-looking and rigorous industry structure.
- Measures at reducing on-farm and off-farm costs and encouraging improved rural sector efficiency.
- Tackling adverse developments in the demand for specific rural commodities.
- Addressing the welfare problems of the rural sector through a wide range of measures. (derived from Commonwealth of Australia, 1986:7).

While all of these points deserve attention, the third point concerning the encouragement of rural sector efficiency is the most relevant to this chapter and will now be examined more thoroughly.

The White Paper argued essentially that the problem of rising costs and falling incomes were primarily the result of depressed world commodity prices, corruption of markets through protection of overseas producers and poor economic management by past governments. The White Paper's key point of significance, however, lay in its claim that governments could do little to address these problems, apart from improving marketing efficiency and removing policy measures that were argued to distort farmer decision-making. As the White Paper noted:

> Many of the problems facing rural Australia today are beyond the capacity of any government to resolve directly. Indeed, for the most part, the adjustment needed to cope with the harsh economic realities that we now face will be undertaken by individual rural producers throughout the country... The Government can play only a catalytic role in bringing about many of these changes. (Commonwealth of Australia, 1986:8)

The significance of this statement is that it emphasized the self-regulatory capacities of farmers as the most appropriate site of intervention for the addressing of on-farm problems. Here, the claimed incapacity of governments to properly intervene was linked directly to attempts to re-constitute the terrain of governing as one focused on the enhancement of individual managerial capacities.

Of further significance in the White Paper was the argument that the farm sector should be seen as part of the broader economic policy-making environment rather than as a special-case. In other words, industry-specific polices that sought to 'prop-up' particular commodity groups in time of stress were regarded no longer as desirable (see Malcolm, 1989). Emphasis was placed particularly on the need to see short-term difficulties as symptoms of longer-term problems associated with farmers' managerial abilities. Strategies that sought to shape the attitudes of farmers and foster a 'prudent' approach to farm management were thus constructed as the only means through which both agricultural productivity and the viability of farmers could appropriately be addressed.

As part of the White Paper, the Commonwealth Government committed an extra $20 million in the form of interest subsidies – introduced in 1985 for farmers assessed as viable in the longer term – to the Rural Adjustment Scheme (RAS). This money was targeted at the debt reconstruction, farm build-up and farm improvement elements of the RAS. The extra funding for interest subsidies was intended to ease the pressures for farm adjustment at a time of high interest rates, rising indebtedness and declining productivity (Commonwealth of Australia, 1986:39). While the release of the White Paper sought to question the role of government in enhancing economic growth in the farm sector, a statement made by the Federal Treasurer one month later criticised fundamentally previous forms of governing, and enabled the attitudes of farmers to assume increased political significance.

The Treasurer's 'Banana Republic' Statement (May 1986)

The key shift in governmental rationalities towards an 'entrepreneurial' and business-focused approach to managing agricultural change occurred during a now famous radio interview conducted on May 14 1986 when the Labor Treasurer, Paul Keating, stated:

> If this government cannot get the adjustment, get manufacturing going again and keep moderate wage outcomes and a sensible economic policy, then Australia is basically done

> for… We will just end up being a third-rate economy…a banana republic. (cited in Kelly, 1992:212)

This important statement, which coincided with a collapse in the balance of payments, foreshadowed a fundamental reordering in governing that was to have major implications for how the practices of farmers were constituted. As Kelly (1992:196) notes, this statement was a turning point that marked a distinctive shift in the Australian policy tradition. Acceptance by the Treasurer, Paul Keating, of advice from his senior advisers that a 'tough-minded treasury-inspired economic rationalist direction…would solve Australia's problems' (Kelly, 1992:204; see also Catley, 1996) constituted the rationality of governing that underpinned this shift.

However, a program of reform would not have been possible without the growing numerical dominance of the Labor Right faction within the Cabinet, and its support by the Centre Left faction. While having no clear majority of votes, the Right was able to frequently muster support to achieve the necessary numbers (Catley, 1996:68). The Left faction became increasingly marginalised following Keating's Banana Republic statement. When, at the 1986 national Labor Party Conference in Hobart, the Left called for a Keynesian-style increase in spending to stimulate economic growth, its proposals were defeated resoundingly by the Labor Right faction. From this point onward, the Left had increasingly diminished influence within the Party and rarely had the numbers to oppose the Right's concern with the promotion of international competition and increases in productive capacity (Catley, 1996:69). The role of the Labor Right was thus significant in enabling Keating's desire for economic reform to become a platform of the entire party.

Keating's Banana Republic statement signalled a concern with enhancing international competitiveness and economic efficiency. As Martin (1990) notes, the Structural Adjustment Committee – a Cabinet sub-committee of senior economic ministers formed following the 1987 Federal election – played a central role in transforming these concerns into distinct policy problems. In this way, the Structural Adjustment Committee can be seen as a key authority in constructing and ordering problems, enrolling other actors, and in transforming such problems into definable policy issues worthy of attention and action. The reforms announced in 1988 to economic and rural adjustment policy (discussed below) represented part of the Structural Adjustment Committee's attempts to constitute the problems and objects of reform.

The Shift Toward Farmer Attitudes as a Site of Reform

Review of the 1985 Rural Adjustment Scheme (March 1988)

It was the review of the 1985 Rural Adjustment Scheme – commissioned in 1987 and released in March 1988 – that first sought to suggest a link between farm viability, and the individual attitudes and skills of farmers themselves. The Review, conducted by Coopers and Lybrand WD Scott, was framed in terms of both the 'structural' economic crisis identified by Treasurer Paul Keating in 1986, and the claimed 'considerable uncertainty' arising from the international economy, restrictive overseas agricultural policies, and domestic policies of restructuring (Coopers and Lybrand WD Scott, 1988:11). In particular, the uncertainty of the changed financial environment, as a result of deregulation of the financial system and the floating of the Australian dollar, was argued to require more rigorous financial management skills. For instance, as the Review noted, 'There will be increased emphasis on technical knowledge and an increased need for market information and for sound management decisions based on that information' (Coopers and Lybrand WD Scott, 1988:15). Only through improved managerial skills was it believed that producers could remain efficient and productive, and thus have the proper capacities to compete in a deregulated economic environment.

However, not all producers were deemed to have the capacities to be good managers. According to the Review authors, for those farmers who were 'in deep financial difficulties', any training was 'too late' and they should leave the industry 'in their own interests and those of the industry' so that their resources could be more productively deployed (Coopers and Lybrand WD Scott, 1988:14). Therefore, restructuring was beginning to be defined according to individual producers' abilities to be either financially responsible or non-responsible risk managers, something emphasized particularly following the introduction of RAS 1993 (see Chapter Four). This represented an early attempt to constitute farmers' capacities as either 'active citizens' or 'targeted populations' (see Dean, 1997, 1999), where only the former were deemed to have the capability of achieving and maintaining viability.

The Review authors agreed with the objective of RAS 1985, which was:

> ...to assist rural industry structural adjustment and to ease adjustment pressure through the provision of adjustment assistance to individual farmers who [we]re considered to have the capacity to achieve and maintain a commercially viable farm business enterprise (Commonwealth of Australia, 1985a),

However, a key criticism was that it gave the misleading intention that its purpose was 'to keep farmers on the land and to subsidise interest payments, without reference to adjustment effects' (Coopers and Lybrand WD Scott, 1988:17). It was argued that the Scheme needed to be revised by placing more emphasis on long-term adjustment and financial management by farmers themselves. The purpose of this was to 'assist in maintaining and improving the efficiency of Australian agriculture and so better place the industry to meet international competition and contribute to the national economy' (Coopers and Lybrand WD Scott, 1988:20). This rationality underpinning the Rural Adjustment Scheme is substantially different from the 1985 RAS (see Chapter Two). In particular, the terms 'efficiency' and 'international competitiveness' appear in place of 'assistance', arguably a key shift in reflection on how governing should take place. Moreover, the claimed necessity for financial management by farmers is linked closely to a construction of the national economy as vulnerable to international competitive pressures (see Beeson and Firth, 1998).

The key recommendations of the RAS Review included:

- The maintenance of interest rate subsidies or loans – introduced originally in 1985 – targeted at farmers for the purposes of debt restructuring, capital investment, purchasing of holdings from farmers considered to be unviable, and carry-on finance for severe short-term market downturns.
- Expansion of adjustment assistance in Part A of the Scheme to include training grants for the improvement of financial and managerial skills, and the provision of

funding for advice and information on technology, training and appropriate farm programs.
- Enhanced incentives for those wishing to leave the industry via the replacement of the rehabilitation loan with a re-establishment grant, reduction in the duration of household support for those taking steps to leave the sector, and the provision of funding for professional financial advice, counselling and information on alternative employment opportunities. (derived from Coopers and Lybrand WD Scott, 1988:28-29)

Two key shifts can be ascertained from the Review of RAS 1985. First, and most obvious, was the increased emphasis on business management skills and the claimed need for formal training in order for farmers to minimise individually the uncertainty arising from external sources of risk. Second, much greater incentives were recommended for those wishing to leave the sector. Not only was it suggested that the re-establishment grant be increased, but also that those provided with household support be given clear signals that a decision to move off their land was necessary as soon as possible. Here, state support was argued to induce dependency and prevent active self-management practices – what scholars such as Rose (1993, 1999) and Dean (1999) refer to as an 'advanced liberal' form of governing.

The Review of RAS 1985 by Coopers and Lybrand WD Scott contributed significantly to a revised Rural Adjustment Scheme, which commenced in January 1989. However, two other policy statements during 1988 were also crucial in shaping the political rationality that constituted the 1989 RAS. These included the Treasurer's *Economic Statement* (Keating, 1988) in May – which emphasized the need for major reforms to industry assistance as well as a re-thinking of the role of government in economic matters – and the release of *Policies for Growth* (Kerin and Cook, 1988), that sought to apply the May Economic Statement to Australian agricultural industries.

Treasurer's Economic Statement (May 1988)

The *Economic Statement* contained major changes in thinking about economy and the forms of intervention that governments could legitimately pursue. It was framed discursively in terms of the claimed need to both reduce the budget deficit and to respond to Australia's balance of payments problems. Changes by government to the financial system and wages system were believed to have only little benefit unless individuals and firms saw also the impetus for change. In other words, a change in attitude was seen to be required. This discourse of attitudinal change was to be again emphasised in the 1991 White Paper *Building a Competitive Australia* (outlined in Chapter Four).

Internationalization was argued to be a necessary part of modernizing the Australian economy, with forms of government regulation seeking to provide support to specific industries argued to be no longer relevant. As Keating (1988:16) noted:

> The way forward for Australia is not to be closeted and sheltered, but to be open and dynamic, trading aggressively in the world. Only this kind of economy can provide the employment and rising living standards that Australians aspire to... Accordingly, the Government will peel away more of this protection.

This statement seeks to constitute a particular model of economy as the *only* credible option. Forms of (Keynesian) regulation based on a discrete national economy are seen as unprogressive, and therefore undesirable, while 'open' economies are constructed as releasing citizens from the shackles of these past evils and producing the desired economic progress. In this sense, the metaphor of protection as being 'peeled away' was very powerful in the constitution of previous forms of governing as misguided which, through institutional reform and attitudinal change, could be rectified to accord with apparent economic realities.

The government argued that an 'outward-looking' and less protected industry policy represented 'the single most effective means of securing greater responsiveness to changing economic conditions and opportunities' (Keating, 1988:119). Industries were expected to increasingly show themselves capable of competing without assistance. The focus on industry adjustment was, perhaps not surprisingly, translated into greater funding for the Rural Adjustment Scheme. This defied the general trend in funding for most agriculturally-based programs. Keating's *Economic Statement* sought to constitute both a model of economy, and a rationality for governing through which subsequent reforms were rendered intelligible. The linking of a program of reform with the viability of farmers was outlined in a more comprehensive manner in a statement by John Kerin (Minister for Primary Industries and Energy) and Peter Cook (Minister for Resources). This statement will now be examined in detail.

Primary Industries and Energy: Policies for Growth (May 1988)

The *Policies for Growth* White Paper was launched shortly after the release of the May *Economic Statement*, and sought to provide details of proposed reform as it applied to the agricultural and resources sectors. This document was released at a time of declining terms-of-trade, depreciation of the Australian dollar and continued subsidization of overseas producers, and it sought to constitute these problems, and solutions to them, in a very specific way. For instance, the objectives of improved competitiveness, enhanced productive capacity and economic efficiency on the part of individuals and firms were argued to be essential in maintaining Australian living standards. Equally, the addressing of economic problems was believed to require not short-term palliatives, but long-term structural change and adjustment. As the document notes:

> The changes that are required are fundamental to the nature of our society and industrial structure. Our industries must become more responsive, diversified and competitive. We must provide the environment for those industries that can make the greatest contribution to develop and grow… The structure of the Australian economy needs to adjust to the market opportunities that emerge. (Kerin and Cook, 1988:2)

Of key interest in the above statement was the argument that investment needed to be attracted to the most productive activities. This represented a shift of emphasis from marginally-viable farmers who it was assumed would know whether or not they were viable, to those producers deemed to have the managerial *capacities* for viability. It was not until 1988 that a discourse of farm management sought to provide a more 'rational' means of differentiating between productive and non-productive farmers. This emerged most clearly in the 1989 version of the RAS, examined later in this chapter. As part of the increased emphasis on managerial skills, a number of changes were announced to the structure of the Rural Adjustment Scheme to make it more effective and increase the often 'less than desirable' rate of structural change (Kerin and Cook, 1988:61). The specific changes to the RAS are detailed in the next section of this chapter. However, it is here worth examining the rationality upon which the truth claims of these changes were based.

While reductions in assistance levels formed a central aspect of the proposed reforms, state intervention was not to be completely ceased. The attraction of resources to those activities defined as productive, and the encouragement of industry adjustment, was believed to require continued intervention and assistance, but in a 'positive' form that would lead to 'less reliance on government assistance in the longer term, and encourage individuals and firms to develop new opportunities' (Kerin and Cook, 1988:7). In other words, a more active 'advanced liberal' form of intervention and citizenship was proposed, aimed towards equipping individuals with the attitudes and culture to change and become more 'outward-looking'. Thus, intervention was not so much withdrawn but was targeted in new ways to foster an 'active' and 'responsible' approach by farmers to their practices. Greater enterprize on the part of individuals was argued therefore to be essential in effecting improved economic efficiency and competitiveness. As the White Paper noted, 'Protection has...induced an inward-looking management attitude that seeks too much from government and too little from entrepreneurship and individual endeavour' (Kerin and Cook, 1988:32). From this perspective, the encouragement of outward looking attitudes and entrepreneurial forms of conduct were seen to be a logical step to improved economic performance with the individual capacities of farmers constituted as the most appropriate sites for intervention and reform.

Of particular significance in the White Paper was the mention for the first time in agri-political discourse – almost in passing – of self-reliance and individual risk management. While not examined in any detail, individual risk management by industries and producers was seen as essential in increasing both choice and efficiency (Kerin and Cook, 1988:37). This represented an important shift in thinking about how the farming sector should be regulated. Rather than state agencies ensuring the maintenance of farm viability through forms of risk collectivization, this was now to be placed in the hands of individual producers themselves. The role of state agencies was constructed as 'reducing the distortions to risk' (Kerin and Cook, 1988:37) instead of assisting directly in its management.

Nevertheless, while beginning to be associated with the capacities of farmers, self-reliance was not yet constituted as an issue of either farm financial management or environmental responsibility. As argued in Chapter Four, it was not until 1990 with the release of the Drought Policy Review Task Force's Report (Commonwealth of Australia, 1990) that self-reliance was constituted for the first time as a strategic way of managing both farm finances and environmental risk. This report sought to argue that producers should be able to manage most forms of risk as part of their normal business activities.

Revised Rural Adjustment Scheme (December 1988)

As noted above, the RAS was restructured extensively in late 1988 following a review of the Scheme in March (Coopers and Lybrand WD Scott, 1988), and the release of the two White Papers (Keating, 1988; Kerin and Cook, 1988). The Review was particularly influential in shaping the revised RAS, with most of its recommendations included in the *States and Northern Territory Grants (Rural Adjustment) Act* (Commonwealth of Australia, 1988a) assented to on the 12th of December 1988. Similar to the previous documents examined, governments were constructed as having progressively less control over the economic fortunes of agriculture. This was seen to necessitate much more 'flexible' policy responses and administrative arrangements. According to the Labor Minister for Primary Industries and Energy, John Kerin, during his second reading speech:

> Adjustment has become an increasingly more significant issue for the farm business because of the greater degree of volatility of external factors it is now likely to face. This volatility arises from a host of circumstances including variability in world commodity prices, exchange rates, interest rates and climatic conditions. Rigid policy responses and smothering administrative regulations cannot cope with this type of variability. (Commonwealth of Australia, 1988b:1462)

Kerin argued that particular forms of state regulation stifled the capacity of industries to compete effectively in an increasingly variable economic environment. More significantly, Kerin constituted the regulatory arrangements of state agencies as part of the problem. For this to be remedied the Australian States were given greater managerial and financial responsibility concerning the everyday operation of the Scheme. While the Australian Constitution has historically obliged the Commonwealth to involve the States in agricultural regulation, a discourse of 'flexibility' and 'responsibility' sought to ensure that State authorities assessed applicants in a manner more consistent with national objectives. In other words, the calculative capacities of State authorities were now to be placed under increased scrutiny by the Federal government to ensure that the most 'rational' decisions were made in assessing applicants. I return to this issue in more detail in the next section of this chapter.

The 1989 RAS contained an emphasis upon building the entrepreneurial capacities of farmers. In this sense, the rationality underpinning the RAS was decidedly advanced liberal with state agencies playing a central role in the building of farmers' managerial capacities. According to the Act the purpose of the Scheme was to 'assist in maintaining and improving the efficiency of Australian rural industry and so better place the industry to meet international competition and contribute to the national economy'. This is the definition recommended by the authors of the RAS Review. Also adopted in the new RAS were the Review's recommendations on the claimed need for adoption by producers of more effective business management skills. The RAS objective can be illustrated by noting some of the comments made by the Minister for Primary Industries and Energy, John Kerin who, during the second reading of the Bill, stated:

> ...there is to be a much stronger emphasis on the notion that farming is first and foremost a *business* – which must be run along sound business lines if it is to survive and prosper...In addition, provision is made for ongoing funding to improve the quality of financial analysis which is undertaken before assistance is provided to the farmer. (Commonwealth of Australia, 1988b:1462, italics V.H.)

This statement marked a decisive shift in governmental rationalities from previous rural adjustment policies in that it encouraged farming to be seen solely as a business enterprise. The 'lifestyle' dimensions of farming were here regarded as secondary to business objectives with successful farming seen as the result of effective business management strategies. In previous policies attachment to farming as a lifestyle was not seen to conflict with farm productivity. Now, however, it was constituted as incompatible and, in fact, was argued to distort 'proper' farm decision-making. Note the following quote from John Kerin: 'The retention of farming as a desirable lifestyle will be a result of effective management of farm businesses and effective adjustment, *not* an objective of it' (Commonwealth of Australia, 1988b:1465, italics V.H.). Such an emphasis on farming as a business enterprise is advanced liberal in that it attempts to de-socialize farm practice and constitute economic efficiency as the truthful end point of conduct (Rose, 1996). Also of note in the above statement is the focus on formal technologies of financial management for farmers to 'prove' their viability in a rational manner, and thereby their capacity to compete in the international marketplace.

The 1989 RAS took the following format:

- A Farm Financial Management Skills Program was added to Part A to provide for improvements in farmers' managerial and technical skills.
- Drought assistance was added to the carry-on assistance provision in Part B of the Scheme, coinciding with the removal of drought from the register of national disasters. A revised assessment criterion allowed assistance to be provided in circumstances where farmers faced financial difficulties due to circumstances beyond their control. Burdon (1996) notes that this represented a crucial change in that assistance could now be provided even if farmers were eligible for funds from normal commercial sources.
- The rehabilitation grant was replaced with an increased re-establishment grant ($28,000 grant indexed to the Consumer Price Index) under Part C to be made available to eligible farmers leaving the industry. The duration of Household Support assistance was reduced from three years to two.

However, despite the Bill being endorsed by all the major parties, some reservations were expressed concerning its capacity to ease the adjustment process for farm families. For example, during the second reading Bruce Lloyd, National Party Member for Murray (New South Wales), noted that: 'There is a need to accommodate and assist this adjustment process so that it is not as brutal for the individuals caught in it' (Commonwealth of Australia, 1988b:2713). Nevertheless, the Member agreed that farm adjustment was a desirable process that should be allowed to continue because it ensures 'that our farmers remain world competitive and our ability to trade is maintained' (Commonwealth of Australia, 1988b:2713). These comments are notable considering the traditional concern of the National Party to maintain farmers on their land. The Liberal Member for Wakefield (South Australia), John Andrew, in opposition to many others, questioned the idea that improved business management would lead to more rational conduct by farmers:

> ...the problem facing many farmers around Australia and the acute problem currently being faced for the third year in a row by drought-stricken farmers on the west coast of South Australia are that their difficulties could not have been identified before assistance was provided. Their problems are not self-inflicted. Their problems are not corrected...by some sort of innovative bookkeeping technique. Their problems are the problems of drought. (Commonwealth of Australia, 1988b:2723)

This statement sought to contest advanced liberal forms of governing agricultural change by raising the problem of events that might not be manageable and foreseeable by farmers. Here, restructuring was constituted as an issue seen to require collectivized forms of regulation and security, rather than a focus on the individual managerial capacities of farmers. Nevertheless, a restructured RAS with a much greater emphasis on business skills, was endorsed by the majority of Members as the only rational option for improving the competitiveness of Australian agriculture by sending the 'correct' signals to farmers, and re-distributing resources to those who would use them most efficiently.

GOVERNING ADMINSTRATIVE FREEDOM THROUGH FINANCIAL ACCOUNTABILITY

As stated in Chapter One and Two, the Australian States play a crucial role in the administration of programs for managing agricultural change, such as the RAS. From 1988, the Commonwealth sought to increase the flexibility with which State adjustment agencies made their assessment of RAS applicants with policy development determined by Ministerial guidelines rather than legislative amendment (Musgrave, 1990). While the tests of eligibility for RAS programs were determined up until this time according to loosely-defined criteria, the Commonwealth had set relatively strict guidelines on how each of the programs within the RAS was to be administered. On the face of it, RAS 1988 seems then to have represented a shift from more direct administrative control by the Commonwealth towards increased power and flexibility by the States.

However, such 'flexibility', in fact, required forms of governing that rendered the conduct of State authorities amenable to surveillance by Commonwealth agencies. For example, administrative functions were framed within stricter financial monitoring guidelines, a trend continued during the 1990s (see Chapter Four). Thus, while the Commonwealth remained obliged legally to achieve its objectives through State authorities, this relationship was re-configured by technologies of governing that ensured authorities would make the most economically rational decisions, but without having their formal decision-making autonomy threatened. Governing through accountability is a good example of what Dean (1999:168-170) refers to as 'technologies of performance' in which formal techniques of calculation exercise scrutiny over the conduct of authorities.

The 1988 review of the RAS by Coopers and Lybrand WD Scott (1988) argued that the administration of RAS 1985 had been 'pedestrian' and that much greater effectiveness and responsiveness was needed. Specifically, the 1985 agreement:

- Did not enable RAS Authorities to address adjustment issues early enough.
- Was too narrow in that it did not allow RAS Authorities to address the broader facets of adjustment such as advisory services and management training.
- Hampered the adjustment process by the requirement that the rehabilitation grant be paid only after the farmer left their property.
- Was unduly restrictive in the use by the RAS Authorities of RAS monies.
- Lacked clarity in linking the instruments to the target for which they were intended.

- Lacked a concise definition of who was eligible for RAS, and thus left too much scope for judgement by State RAS Authorities. (derived from Coopers and Lybrand WD Scott, 1988:18-19).

On this basis the Review authors recommended that the State authorities needed to be given increased powers to manage RAS funding in their States. However, this claimed increase in power was to be 'accompanied by the acceptance of responsibility (far greater than in the past) by the States for accountability under the RAS' (Coopers and Lybrand WD Scott, 1988:32). In other words, while it was believed that the States required administrative freedom in order to respond to specific local and regional conditions, this flexibility was governed via technologies of performance, specifically financial responsibility and accountability.

The Commonwealth accepted the Review's findings and sought to devolve greater flexibility to the States. According to the Minister for Primary Industries and Energy, John Kerin, to provide for this, the RAS 1989 legislation and agreement:

- contain a clear mission statement for the operation of the Rural Adjustment Scheme;
- provide for the State rural adjustment authority to have sole responsibility for the management and administration of the policy measures available under the scheme;
- provide for the State rural adjustment authorities to have total freedom, within the bounds of responsible financial administration, to manage funds provided for the purpose of the scheme; and
- provide for the establishment of a management information system to meet the States' own management requirements and to allow them to report to the Commonwealth on the effectiveness of the Scheme's operation' (Commonwealth of Australia, 1988b:1463).

The principal functions of the State authorities were now defined as: 1) assessing the eligibility of farmers for assistance; 2) determining the most appropriate form of assistance; 3) delivering funds and services in accordance with Commonwealth policy guidelines; 4) reviewing the effectiveness of assistance on a periodic basis, and 5) promoting the scheme to the farming and financial sectors (Commonwealth of Australia, 1988a:s.7(5)). A central aspect of the new arrangements was the increased emphasis on the (contractual) obligation of State administering bodies to be financially responsible and accountable. State administrations were expected to use their freedom in a financially prudent manner. For example, as Minister Kerin noted:

> I cannot stress too much this requirement for accountability. A State which misuses this freedom, for example, to distort the scheme's objectives or to shift the risk of other State activities on to the scheme, will lose access to Commonwealth funds (Commonwealth of Australia, 1988b:1464).

This statement shows how the Commonwealth sought to achieve its objectives through financial accountability standards. These standards represented a technological means of aligning the conduct of State administrative bodies with the advanced liberal rationalities of governing underpinning Commonwealth objectives.

Despite the emphasis on greater financial accountability, it appears that in practice the procedures of State administrative bodies changed little. As a Commonwealth public servant interviewed in the course of the research for this book noted, this has had much to do with the level of monitoring required for a State agency to reasonably translate Commonwealth objectives:

> [The] determination of what category under RAS the money was going to be spent on at the end of the day was a Commonwealth decision and not a State decision. But having made the decision…and having developed general policy guidelines, it was then to the States to interpret the guidelines and run the scheme on a day-to-day basis and that is where you have complete freedom. We just don't have the resources to be able to check every one of those decisions… The State could make a decision which it thought was entirely practicable and so forth and we might take a different view, but the day-to-day running of the scheme definitely rests with the states and they are responsible for the judgment…supposedly consistent with the guidelines we set. (Male, Branch Director level [1], DPIE)

Another Commonwealth public servant echoed a similar opinion:

> I think the Commonwealth could have done more to promote the Scheme and clear the way, and to ensure that the States were informed. I guess it does get into the realm where…you're relying on another jurisdiction to do the administration. You don't have to take control and, in fact, I think deliberately the Commonwealth hasn't taken more control because there are benefits in having the State agencies involved and they're closer to the stakeholders. (Male, Middle Management level, RIRDC)

The above statements reflect the nature of liberal forms of rule in which govern*mentalities* are 'eternally optimistic', yet the practice of government is 'congenitally failing' (Miller and Rose, 1990). In the case of managing agricultural change, State authorities are seen as both necessary (under the Federal Constitution) for the achievement of Commonwealth objectives, as well as the key agents who have the capacity to compromise these objectives.

The Mid-Term Review of RAS 1993, released in 1997, provides further evidence of the problems of governing State administrations in a Federal system of government. This Review noted that while Commonwealth guidelines were generally followed, State administrations were applying further subjective criteria in making their assessments of RAS applicants (McColl et al., 1997). In other words, State administrative bodies were 'translating' (Callon, 1986; Rose, 1999) Commonwealth objectives in ways consistent with their own understandings of farm viability. As one Queensland public servant noted, this had much to do with the inherent difficulties in standardizing interpretations:

> …the various State's roles is an interesting one, because up until probably the early nineties there was a hell of a lot of difference between the way in which the States administered RAS, and I dare to say, that if you put all the States next to each other, everyone would have a slightly different interpretation of the guidelines and how to apply and so on. (Male, Middle Management level, Queensland Treasury)

Another Queensland public servant was more specific noting that the key difficulty in achieving greater standardization was the problem of defining viability:

> ...the most difficult part of the Rural Adjustment Scheme is the ambiguity of concepts like viability – how do you determine who is viable and who is not? Are they the people that are at the extreme ends of the scale? And if they are eligible for the Rural Adjustment Scheme you assume that they are temporarily unviable but some time down the track they will become viable. So that is the central policy difficulty of that scheme, and often nobody is happy, you know. (Male, Senior Officer level, Qld. Office of Rural Communities)

Despite criticisms that the States did not conform closely enough to the objectives of the RAS, the Commonwealth continued to endorse flexibility. Clearly this allowed State authorities to be sensitive to the individual problems of farmers. Fundamentally, however, it enabled also technologies of performance to be deployed by Commonwealth agencies, such as accounting standards, through which assessment procedures could be monitored in a more economically calculable way than retrospective legislative amendments. Nevertheless, it appears that the problems with defining farm viability, and the regional differences between State administrations, rendered the interpretation of Commonwealth objectives subject to continuing problems. Many Commonwealth public servants appear to accept, albeit reluctantly, that the continuing diversity of interpretations is necessary for farm adjustment programs to operate in a Federal system of government, and that stricter monitoring might result in undue levels of resistance by State authorities. This supports O'Malley's (1996) argument that the original objectives of programs are frequently compromised from the outset due to the need to negotiate concessions that may be necessary to achieve the desired result.

Creating More 'Responsive' Rural Industries: The National Farmers' Federation

The period from 1986 to 1989 was one where the National Farmers' Federation (NFF) assumed an increasing input into formal political life, and sought to convince the Federal government of the need for widespread economic reform to avert the terms-of-trade decline. This was a time when farm organizations in general were giving more sustained attention to professional input into government policy-making (Malcolm, 1989; Trebeck, 1990). However, it was the NFF who adopted the most visible, and perhaps controversial, political profile through their anti-union stance, and support for agricultural efficiency. In fact, from 1986 the NFF promoted itself as the voice of rural Australia, thereby challenging the previously unquestioned role of the National Party in defining rural policy issues and problems (see Connors, 1996).

The exposure of Australian industry to market forces, the creation of an internationally competitive economy, a de-regulated labour market, and freer trade constituted the political rationality that united the NFF (Trebeck, 1990; Kelly, 1992). Its commitment to the removal of protectionist policies placed it in direct opposition to the National Party. The NFF sought to position itself against the perceived 'agrarian socialism' of the National Party, and the regulated labour policies of the Labor Party, by espousing an aggressive market-driven stance on micro-economic reform. As the peak Australian farming organization, the NFF could claim to speak on behalf of all farmers, and thus argue that its policies were essential to the economic well-being of Australian agriculture.

The NFF's objectives for reform are contained in its 1986 White Paper on *Economic and Rural Policy* (National Farmers' Federation, 1986). While primarily a response to the Commonwealth White Paper *Economic and Rural Policy*, released in April 1986, the NFF document proposed much more radical measures to change the attitudes of primary producers. Many of their recommendations were later echoed in the 1988 White Paper *Primary Industries and Energy: Policies for Growth* (Kerin and Cook, 1988). The NFF White Paper argued that short-term assistance packages, and policies of industry protection, would do little in the longer term to address Australia's economic problems. Major economic reforms were seen to be required to reduce both off-farm and on-farm costs through the reduction of claimed structural impediments, instead of through subsidization and protection – which were argued to shield farmers from market forces. As the document states:

> What is required is a major shift in the thrust of Australia's macroeconomic policy aimed at achieving lasting reductions in the rate of inflation, significant declines in interest rates, improving our trade performance and allowing adequate adjustment to our declining terms of trade. Such a policy shift must be accompanied by other industry specific measures directed at making Australian industries more outward looking and responsive to the demands of the international marketplace. (National Farmers' Federation, 1986:1)

These changes were argued by the Paper to be of benefit to not only the farm sector but to all Australians. Hence, industry protection and introspective attitudes were constructed as inhibiting Australia's economic performance, and providing benefits to only a small section of the population. Economy-wide reform through wages restraint, increased productivity, and reductions in protection were argued to allow adjustment to market realities, aid wealth creation (rather than distribution) and benefit all in the process. The NFF was here attempting to construct past forms of governing as concealing the true dynamics on which economic life operates. Thus, the NFF constituted a meta-narrative through which reform could be thought, and acted upon. Reform was seen as inevitable, and was argued to be possible only through forms of regulation that transformed individual attitudes: an advanced liberal form of governing. This way of reflecting on rule had clear links with the political rationality that was emerging in Commonwealth government documents from 1986. While many of the NFF's objectives remained unpopular with the governing Labor Party of the day – particularly its dislike of the trade unions – many of its objectives show striking similarities with those of the 1988 White Papers (ie. Keating, 1988; Kerin and Cook, 1988), examined in an earlier section of this chapter.

CONCLUSION

Agricultural restructuring was constituted from 1986 according to an advanced liberal rationality of governing that focused increasingly on individual farm business capacities as the key to a more efficient economy. This involved a questioning of the Australian economy as a self-contained unit, and the constitution of individual business capacities as inseparable from the claimed imperative of international competition. Generally, the national economy during this period was constructed as in crisis with previous forms of governing seen as exacerbating this problem. Authorities sought to argue that through the facilitation of outward looking attitudes, and greater entrepreneurialism, farmers would have enhanced capacities to compete efficiently in international markets. The attitudes and conduct of farmers thus became inseparable from how the national economy was defined. Such a rationality was endorsed strongly by the National Farmers' Federation, the peak Australian farming organization, whose economic reform objectives carried increasing political weight as the voice of rural Australia. In the case of programs for managing change, such as the RAS, the claimed imperative for economic restructuring and a less interventionist political approach was reflected in a stronger focus on the adoption of formal farm business skills, as well as increased monitoring of State authorities through technologies of performance.

These events indicate the emergence of an advanced liberal way of reflecting on rule in which the self-regulatory capacities of individuals were constituted as the most appropriate sites of reform. RAS 1989 was particularly reflective of this rationality of governing in that it encouraged formal business skills as essential in running a viable farm. Thus, business management skills now assumed a central status as the 'benchmark' through which the conduct of farmers as productive or non-productive could be defined and assessed. Nevertheless, it is important also to acknowledge the continued problems in making this distinction at the State level, and the ongoing tensions between Commonwealth objectives and their practice by State authorities, a point re-visited in Chapter Four.

The next chapter explores how the managerial capacities of farmers assumed further significance during the 1990s through the notions of 'self-reliance' and 'individual risk management'. While a concern with international competitiveness was retained, the business management skills of farmers were constituted increasingly as a key site of risk, and therefore restructuring. Formal training in business management skills emerged through the 1990s as a central means by which farmers could 'properly' manage financial and climatic change, and thus become 'active' managers of themselves and their farms.

Chapter 4

TOWARDS SELF-RELIANCE: GOVERNING THE FARMER AS INDIVIDUAL RISK MANAGER

Since the early 1990s, farmer 'self-reliance' has received widespread support at a government and non-government level throughout Australia (eg. Queensland Government, 1992; Synapse Consulting, 1992; National Farmers' Federation, 1993; Commonwealth of Australia, 1994; McColl et al., 1997) with primary producers encouraged, through various policy instruments, to manage, in increasingly 'entrepreneurial' ways, financial and environmental risks. On the one hand, self-reliance is argued in official political documents to comprise an essential part of farm management, leading to both profitability and long-term environmental sustainability. On the other hand, those farmers regarded by authorities as not having the capacities to become self-reliant are deemed as hindering the development of a more internationally competitive sector, and are argued to have a limited future in primary production.

While it is tempting to interpret the emergence of self-reliance in political discourse as a neutral means of improving some farmers' managerial capabilities, at the same time as facilitating the exit of other producers, the genealogy of governing developed in previous chapters suggests more caution. Rather than simply an inevitable response to improving agricultural efficiency and sustainability, self-reliance represents an assemblage of governmental rationalities and technologies that seek to govern farmer conduct in particular ways. Thus, programs that encourage self-reliance seem to represent a desire on the part of authorities to govern through the personal calculative capacities of individuals: what scholars such as Rose (1993, 1999) refer to as an *advanced liberal* form of rule. This chapter explores how a self-reliant approach to managing agricultural change in an advanced liberal way was assembled between 1990 and 1997 as the most truthful means of dealing with financial and environmental risk. It focuses particularly on how this strategy for restructuring extended the earlier emphasis of international competitiveness through improving farmers' managerial capacities.

ENVIRONMENTAL SUSTAINABILITY: A FINANCIAL MANAGEMENT PROBLEM

Self-reliance emerged in the early 1990s as a strategy that aimed to improve both farm financial management (and thus profitability), and contribute to a more sustainable agriculture. In order to appreciate how both environmental and economic concerns were constituted as a unified problem, and assembled into a programmatic form for governing the conduct of farmers, the concept of *sustainable development*, which gained political currency during the mid-1980s (Murdoch and Clark, 1994), needs to be briefly considered.

It is first important to note that *sustainability*, or *sustainable development*, are not self-evident or conceptually unproblematic terms. As Redclift (1993) argues, sustainable development has economic, socio-political and epistemological dimensions that influence how it is constituted as an object of knowledge. From the perspective of a genealogy of government developed in this book, 'sustainability' needs to be seen as an effect of how governing is reflected upon as a practical exercise. This means that it is inappropriate to reduce sustainable development, and sustainability, to either an end-point that can be achieved through the application of scientific knowledge, or to an ideology that legitimates the claimed logic of an expanding capitalist system. Rather, it should be represented as an ongoing site of struggle in which 'sustainable knowledge' is constituted through heterogenous practices (Murdoch and Clark, 1994).

Since European settlement, the natural environment in Australia has been constituted at a formal political level in terms of its productive capacity and potential to generate profit (Dovers, 1992). While agricultural landscapes continue to be defined predominantly in these terms (see Lawrence, 1995), the late 1980s saw a subtle shift in the way in which the natural environment was constituted. Recognition that soil erosion, salinity and land clearing were potential problems (Watson, 1992) led to increasing attention to issues of environmental degradation. This concern was reflected in the announcement in July 1989 by Bob Hawke, the Labor Prime Minister, of a National Landcare Program (Lockie, 1999). However, the apparent commitment to addressing farm environmental problems, embodied in programs such as Landcare, was shaped by the Hawke and, from 1992, the Keating Labor government's desire for national economic reform.

Applied to agriculture, the effect of programs such as Landcare was to construct environmental degradation as an individual risk to be factored into farmers' planning in order for long-term 'sustainable' outcomes to be achieved (Martin, 1997a, 1997b; Lockie, 1999). Thus, environmental problems were rendered knowable in an advanced liberal way: they constituted sources of risk that, if managed in an entrepreneurial manner, would produce long-term financial and productivity benefits.[1] The constitution of environmental sustainability as an individual problem was significant in enabling self-reliance to emerge as a means of enhancing farmers' capacities to manage this risk.

---

[1] Robertson (1997) provides an excellent example of this viewpoint.

The Constitution of Self-Reliance as a Strategy for Risk Management

The construction of sustainability as an individual risk needs to be considered as a key element in the emergence of self-reliance as an advanced liberal form of governing the conduct of farmers. However, the constitution of self-reliance as a means for managing financial and environmental risk would have not been possible without first a shift in rationalities of drought management.

Drought Policy Task Force Review (1990)

Historically, drought in Australia was conceptualized at all levels of government as a 'natural disaster'; a random severe event that could properly be managed by farmers only with public assistance. In other words, the capacity for farmers to manage was seen as dependent on the existence of *public* programs – in particular, the Natural Disaster Relief Arrangements (NDRA) – and relief measures that assisted them to minimise the effects of drought. Rose (1993, 1999) and others refer to this type of intervention as a *welfarist* form of governing. The status of drought as a disaster came to be called into question increasingly during the 1980s, particularly as a result of the level of assistance required during the extensive and severe drought of 1982/83 (Stehlik et al., 1999). These views were reinforced by the Industries Assistance Commission in 1983 who noted that those farmers who had taken steps to prepare for drought were, in effect, penalised by drought relief being directed at those with high debt and few readily available assets. This was argued to reduce incentives for 'farmers to make physical and financial provision for future drought' (Industries Assistance Commission, 1983:13). Such recommendations on drought relief formed the basis of a major inquiry that changed the way in which drought support was delivered.

A Drought Policy Review was commissioned in 1989 by the Hawke Labor Government (Commonwealth of Australia, 1990) in response to the conclusions of the IAC's 1983 report. This report changed fundamentally how drought was defined and governed. The Drought Task Force Report represented one of the first attempts to question the status of drought as an 'external' risk, and outline self-reliance as a strategic means through which farmers could more rationally manage their properties. Importantly, drought in this report was reconstituted from a natural disaster, through the Natural Disaster Relief Arrangements, to something that could be managed for and planned as part of farmers' 'normal' risk management strategies. As the Task Force noted:

> Managing for climate...must become the norm instead of what has amounted in the past to attempted income and climate stabilization measures. The need to manage for variable climatic conditions puts an onus on producers to adopt more flexible farming and management strategies (Commonwealth of Australia, 1990:4).

The Task Force also argued that:

> While drought may be triggered by long periods of rainfall deficiency, it represents a prolonged failure or inability of producers to respond to those deteriorating conditions.

> Extreme drought represents periods of extreme risk to agricultural industries and the rural communities. Drought is therefore a relative concept that reflects the fact that the current agricultural production is out of equilibrium with prevailing seasonal conditions. Managing for drought, then, is about managing for the risks involved in carrying out agricultural business in a variable climate (Commonwealth of Australia, 1990:7).

The significance of this shift in governmental rationalities from drought as 'natural disaster' to 'managed risk' lies in both the changing locus of risk and the imputation of agency to farmers to manage such risk. Drought was brought into an advanced liberal domain of calculation through the linking of the natural environment – defined as an individual risk – with the practices of farmers. It was represented no longer as an event external to the planning strategies of farmers, and manageable through collective regulatory mechanisms, but a material effect of farm management strategies. In other words, the existence or absence of drought was reconfigured within the capacities of farmers to conduct themselves in an entrepreneurial manner. Drought could thus be addressed only by first acting on the capacities of producers and constituting them as *active* agents in the management of risk (see Dean, 1997, 1999). On the one hand, an environmentally responsible farmer was one who had, using the correct techniques of business management, calculated accurately the risks of drought and taken steps to mitigate this risk. On the other hand, the inability of farmers to deal with what they defined as drought was now taken to signify poor management skills that rendered a property 'unsustainable': such farmers were constituted as having a limited future in agriculture.

The consequences of such a shift are not difficult to observe. As a natural disaster, drought was essentially an act of God that could not technically be controlled, although it could be planned for and managed by farmers through government incentives such as Income Equalization Deposits (IEDs) – which enabled a farmer to put income aside in good years for use in times of drought – and the assurance that collective measures were available should the impact of drought significantly affect farm production. When a drought occurred it was expected that state agencies would respond in order to assist the financial and welfare needs of farmers. Once constructed as a normal risk, responsibility was shifted solely to the individual farmer who was expected to conduct him or herself in a financially prudent manner to avert the effects of drought. Drought was to be treated in a similar manner to market fluctuations: a random, yet calculable event, the regularities of which could be more 'truthfully' ascertained and controlled through economic expertise and business training. In addition to financial incentives of preparedness, such as IEDs, the role of state agencies was now primarily to ensure that farmers were equipped with the decision-making capacities to prepare for drought. Drawing upon the above arguments, self-reliance can be seen as being based on an advanced liberal way of reflecting on rule in which authorities seek to transform farmers into active managers of risk via various marketized self-regulatory techniques.

It is important to note that self-reliance, as it is constituted in the statements of the Drought Policy Review Task Force, is very different from its historical agrarian fundamentalist associations with 'independence', 'freedom' and 'being one's own boss' (see Aitkin 1988; Gray 1991; Kidman 1991). Rather than associated with farming as a virtuous and noble way of life, self-reliance was essentially made knowable through discourses of entrepreneurialism, economic efficiency and business management. While claiming to increase personal choice and autonomy – something that sits comfortably with farmers' attachment to independence and hard work (see Lockie, 1999) – self-reliance redefined farm viability in purely commercial terms, and farmers themselves as economically rational actors who, if equipped with the 'right' business skills, could manage risk effectively.

The Task Force report contributed significantly to the National Drought Policy, released in 1992. This policy is important to consider as drought assistance subsequently became linked to the Rural Adjustment Scheme: the central program from 1977 through which agricultural change was governed. However, the release of a major government policy statement in 1991 was important in establishing a competitive rationality of governing in which an individual approach to managing risk could be rendered intelligible.

White Paper: Building a Competitive Australia (1991)

The Drought Policy Review Task Force sought to outline the importance of a self-reliant approach to risk management at the farm level. In 1991, the White Paper *Building a Competitive Australia* attempted to link the objective of individual risk management to a desire for national economic restructuring. A key point in this document was Labor Prime Minister Bob Hawke's argument that Australians needed to adopt practices and attitudes enabling an improvement in national productive capacity (Dept. of the Prime Minister and Cabinet, 1991:1.2). In effect, this statement sought to focus objectives of restructuring on the capacities of individuals. Both longer-term 'structural' change, and a shift in 'deeply entrenched attitudes', were put forward as imperative if Australia was to become a competitive nation. In fact, for the Treasurer, Paul Keating, the discipline of the market was seen as crucial in exposing producers to greater competition and 'renouncing the fallacious doctrine that prosperity can be found behind the insular wall of protection' (Dept. of the Prime Minister and Cabinet, 1991:2.2).

These statements construct less protection, greater competition and a change in attitudes as the *only* way forward. Indeed competition is seen as a historically natural and universal standard that might have been applied in earlier years if only the people then were not so 'inward-looking'. Genealogically, this representation of history needs to be questioned as it constructs competition and enterprize as the truthful standard by which all previous forms of governing should be assessed. In constructing history in such a way, present policies are depoliticized as the culmination of an orderly historical progression toward increasingly sophisticated and efficient forms of governing. While attempting to constitute an economy characterized by competitiveness and economically rational decision-making, the population is constituted also as human capital who are to be equipped with the necessary skills to make them self-reliant and therefore productive (see Beeson and Firth, 1998). Here, the construction of a particular model of economic activity was important for the individual management of risk to be brought into a domain of calculation.

The White Paper *Building a Competitive Australia* was crucial therefore in defining the terrain of governmental activity through which individual conduct could be shaped. A competitive economy was constituted as the basis on which all objectives for governing should be assessed. Although beyond the scope of this book, it should be noted that such a rationality formed the basis of the Hilmer Report on competition published in 1993, and the establishment of a National Competition Policy in 1995 (Beeson and Firth, 1998).

National Drought Policy (1992)

Released in 1992, the National Drought Policy was preceded by a Senate Report on 'appropriate government responses to the recommendations of the Drought Policy Review Task Force' (Commonwealth of Australia, 1992a:vii). The task of the Senate Standing Committee on Rural and Regional Affairs was to assess evidence on this issue and prepare a report on its findings. According to the Committee, there was unanimous agreement that the purpose of a national drought policy should be to facilitate a self-reliant approach to drought. The objective of self-reliance was supported most strongly by producer organizations such as the National Farmers' Federation (NFF), the United Farmers and Stockowners of South Australia, the New South Wales Farmers' Association and the United Graziers Association (UGA). Overall, the Committee endorsed the Report of the Drought Policy Review Task Force, and recommended that the Minister for Primary Industries and Energy, in consultation with members of the Australian Agricultural Council, 'give urgent attention to the development and implementation of a clearly defined and comprehensive approach to managing for drought' (Commonwealth of Australia, 1992a:18).

While the White Paper *Building a Competitive Australia* made it clear that the role of state agencies was to facilitate economic growth by changing attitudes and conduct, the *National Drought Policy*, agreed upon by the Commonwealth and States in 1992, linked self-reliance – as a means of risk management – with 'cultural change' on the part of farmers. It is important to note that this policy was released at a time when much of inland Queensland and New South Wales were experiencing a prolonged dry period. Thus, the Drought Policy was clearly significant in showing that the Commonwealth Government was doing something for farmers in distress. However, the emerging rationality of self-reliance sought to define the role of government as providing incentives for farmers to manage drought personally, rather than through the assistance packages of the past. The objectives of the Drought Policy were to:

- encourage primary producers and other sections of rural Australia to adopt self-reliant approaches to managing for climatic variability;
- maintain and protect Australia's agricultural and environmental resource base during periods of extreme climatic stress; and
- ensure early recovery of agricultural and rural industries, consistent with long-term sustainable levels (Commonwealth of Australia, 1992b:1).

While the Federal government claimed to provide a favourable overall environment for achieving these objectives, farmers were expected to respond in a rational manner by implementing business strategies to manage efficiently both financial and physical resources (Commonwealth of Australia, 1992b:1). Income smoothing arrangements such as tax averaging and Income Equalization Deposits (IEDs) were encouraged particularly in ensuring that producers adopted efficient financial management practices. The role of government was constructed as providing a conducive environment for property planning through such incentives as information transfer and formal education and training. State governments were allowed to provide additional drought assistance measures, but these were to be consistent with Commonwealth objectives. It is here apparent how restructuring was beginning to be defined according to the adoption of formal technologies of business management and accounting (outlined in more detail in later sections of this chapter). It was assumed that those primary producers with formal business skills would be the most prudent and, therefore, self-reliant managers of environmental and financial risk.

Review of the 1989 Rural Adjustment Scheme (1992)

The idea that environmental sustainability could be addressed through a more 'entrepreneurial' approach to farm management was constituted further through the review of RAS 1989, in 1992, by Synapse Consulting. This review claimed that international and domestic restructuring had increased the pressures for adjustment, and required therefore a sector that was 'inherently more adaptable, self-reliant and profitable' (Synapse Consulting, 1992:ii). For farmers, this was seen to necessitate the adoption of an 'entrepreneurial and professional approach' (1992:ii) to risk management. It recommended that the main purpose of any adjustment scheme should be to improve competitiveness and productivity, and have close links with micro-economic reform (which the RAS was apparently not doing). RAS 1989 was criticized in this respect for sending the 'wrong' economic signals to farmers through its references of providing assistance – via interest subsidies – particularly in circumstances considered to be beyond the control of farmers.

Accordingly, this Review argued that the RAS should be replaced with a new National Farm Adjustment Scheme, the purposes of which should be:

- Increased sustainable economic returns from the physical, financial and human resources assigned to the farm unit.
- Improving the performance of those farming operations that contribute significantly to the economic role of the sector, or have the potential to do so.
- A focus on the economic performance of the farm rather than the occupational nature of the farm owner, the interest load borne by the farm or the extent to which causes of poor performance might be judged to be beyond the control of the farmer.
- To achieve sustainable improvement in economic performance, the Scheme should aim to remove the causes rather than the symptoms of low productivity.
- To take into account the industry and regional factors affecting the potential productivity gains of farmers. (derived from Synapse Consulting, 1992:31)

These objectives accord with an advanced liberal rationality of governing in three important ways. First, viability was defined in terms of the commercial efficiency of the entire farm unit, with humans and the physical environment constituted as economic resources that could be managed in a formalized manner to optimize personal profit. Second, the Scheme was to be targeted towards already profitable farms – those who made 'a significant economic contribution' (Synapse Consulting, 1992:31) – implying that certain farmers did not have the capacities for active conduct to improve their economic performance in the longer term. Finally, farm performance was seen as an individual economic risk that needed to be managed in order for viability to be achieved. Thus, poor financial performance was equated with a lack of managerial capacity, and an 'unviable' farming operation.

The Review argued also that the achievement of objectives of productivity and profitability required the creation of two programs within the National Farm Adjustment Scheme: a Farm Adjustment Financing Program and a Farm Adjustment Skills Program. The objective of the Farm Adjustment Financing Program was to 'increase the effectiveness of farm adjustment by improving farm productivity and to assist orderly adjustment by reducing constraints on farmers adjusting out of farm ownership' (Synapse Consulting, 1992:35). Its central aim was to provide those farmers defined as productive with grants through which they could gain the 'appropriate' managerial skills or financial advice so that effective farm financing strategies were developed. On the other hand, re-establishment grants of up to $50,000 were to be offered to provide farmers defined as non-efficient with an 'incentive' to leave the industry.

In a complementary manner, the purpose of the Farm Adjustment Skills Program was to 'improve the financial and risk management skills of persons involved in the distribution and use of funds in the farm sector and to identify constraints to the effective and efficient distribution and use of funds in the farm sector' (Synapse Consulting, 1992:36). The main aim of this program was to provide grants for farmers to improve their risk management practices and in so doing enable farmers to 'better evaluate their needs' (1992:37). Here, self-reliance assumed a distinctly technological form for shaping the agency of farmers. Farmers were to be offered 'choices' from a range of economic options, through which they could evaluate their economic position. Thus the aim of the Farm Adjustment Skills Program was to equip farmers with forms of knowledge and techniques of evaluation to make a self-assessment of their financial circumstances, and thereby make the most 'rational' decisions concerning their future.

While the encouragement of managerial skills formed a central part of the Review's argument, the authors acknowledged reluctantly that only anecdotal evidence from bankers existed to suggest that there was a link between financial management and farm viability (Synapse Consulting, 1992:39). Poor prices and seasons, low levels of capital, and inadequate farm size were seen to be responsible for most of the problems that farmers faced. Nevertheless, the Review considered that 'the productivity and stability of the farm sector would be enhanced if the general principles of agricultural finance and economics were more widely understood' (Synapse Consulting, 1992:40). In effect, the Review here attempted to authorise a discourse through which farmers' conduct could be measured, predicted and thereby judged. Genealogically, this represented an exercise of power that attempted to impose a unitary logic and truth on diverse circumstances. The Synapse Review constituted self-reliance according to advanced liberal 'technologies of agency' (see Dean, 1997, 1999), emphasising the need for individual management of financial and natural resources – according to disciplines of accounting and business management – to achieve the goal of maximizing wealth. In this sense self-reliance was equated with one's individual capacities to be *financially* sustainable, with environmental issues constituted as matters to be addressed through improved financial management skills.

Rural Adjustment Scheme (1993)

The specific programs outlined in the Synapse Review were rejected by the then Labor Government. Nevertheless, its political rationality of 'managed risk' was incorporated in the

Rural Adjustment Act, assented to on the 24th of December 1992, and inscribed in a new Rural Adjustment Scheme that commenced in January 1993. According to Simon Crean, the Minister for Primary Industries and Energy:

> The Government's vision is a farm sector which is able to thrive in a highly variable economic and natural resource environment, without relying on government assistance for survival and growth.... We [The Government] will ensure that the operating environment within which farmers' work is conducted is conducive to sound business decision making, and that the messages farmers are receiving from us are consistent with an emphasis on self-reliance and risk management. (Commonwealth of Australia, 1992c:2412)

For the Commonwealth government, this 'vision' involved the constitution of farmers as business managers, who were seen to require skills and training in order to become not only more 'resilient' and 'competitive', but make the most of their economic opportunities (Commonwealth of Australia, 1992c). Self-reliance was equated with the adoption of business management practices, through which farmers could manage their risks in a profitable manner, without having to earn an off-farm income (which was argued to distort on-farm management practices). Through skills in business management, both the economic and physical environments were constructed as manageable. Such a view assumes that greater economic efficiency is not only rational, but contributes also to environmental protection (see Gow, 1996). RAS 1993, like other programs such as Landcare, was thus advanced liberal, in that it attempted to shape farmers' self-regulatory practices so that their choices were economically 'responsible' and 'rational' (see Martin, 1997b; Lockie, 1999).

The focus in RAS 1993 on the enhancing of farmers' capacities to be self-reliant was emphasized in my interviews with Commonwealth public servants. Many of those interviewed believed that RAS 1993 (referred to by participants as RAS 1992) represented a significant shift in thinking about governing adjustment. For example:

> [RAS] 1992...was the one that identified that they had to shift [towards self-reliance]. Now the others had basically been involved with some sort of shifting, but that was the one that really started the shift (Male, Branch Director level [1], DPIE).
> RAS '88 was mostly about assistance. RAS '92 brought in the concept of self-reliance (Female, Senior Officer level, DPIE).
>
> RAS '92...was the first step about self-reliance. It was saying, 'we expect you to be self-reliant, we expect we will give you business assistance in terms of interest subsidies for productivity improvement and we will give you training assistance. We will provide a mechanism whereby people can actually adjust off their farm. But we will still recognise that there are situations in which no producer could be expected to plan and prepare for'. (Female, Senior Management, QDPI)

These statements draw attention to the emphasis on farmers' managerial capacities that RAS 1993 helped to create, in which previous forms of governing farm practice were called into question, and self-reliance constituted as a means of addressing these problems. Analysis of the relevant legislation shows clearly how rationalities of governing were attempting to question and redefine both farmers' economic environment, and the types of practices that farmers should adopt to best respond to this new set of circumstances. Consistent with the recommendations of the Synapse review of the RAS, the objectives of the *Rural Adjustment Act* were to (a) foster the development of a more profitable farm sector that is able to operate competitively in a deregulated financial and market environment; and (b) improve the competitiveness of the farm sector in a sustainable manner (Commonwealth of Australia, 1992c). The new Scheme contained an explicit focus on productivity improvement and long-term viability by making eligibility for interest subsidies conditional on the farm being able to secure commercial finance. This provision was not part of previous Schemes. The legislation provided also for a Mid-Term Review of the Scheme after three years.

As part of the Act, a Rural Adjustment Scheme Advisory Council (RASAC) was created 'to provide strategic oversight of the Scheme' and ensure that RAS fulfilled its objectives of fostering greater self-reliance and individual risk management (Burdon, 1996:28). The purpose of RASAC was to provide also feedback to the Minister on the operation and effectiveness of the RAS. In this sense, RASAC might be seen as a key 'centre' or 'obligatory point of passage' (Callon, 1986) in ensuring that both the economic conduct of farmers eligible for RAS assistance, and the assessment procedures of State authorities, conformed to objectives of self-reliance. RAS 1993 had five components:

- Skills enhancement measures – grants to eligible farmers to upgrade farm business and property management skills.
- Farm productivity enhancement measures – interest subsidies of up to 50 per cent of the cost of commercial finance to viable farmers for productivity improvement measures.
- Re-establishment provisions – an enhanced grant of up to $45,000 for farmers who wished to leave farming.
- Land trading – the trading of land by a State to speed up the process of amalgamation or retire land no longer suitable for agricultural production.
- Exceptional circumstances – interest-rate subsidies of up to 100 per cent on commercial finance for severe events 'outside normal risk management strategies' and for which farmers could not reasonably be expected to plan. (derived from Rural Adjustment Scheme Advisory Council, 1996:24-25)

While encouraging self-reliance, the Commonwealth (Labor) government attempted also to establish a commitment to social justice as part of this approach. This commitment, however, was constituted in terms of short-term welfare support via the Farm Household Support Scheme, which was removed from the RAS in the 1992 Act. Provision for social support had been a central aspect of the RAS since its inception in 1977. Household support, it appears, was regarded as both distorting the business signals that the RAS was supposed to be sending to farmers, and leading to insufficient 'unviable' producers leaving the industry. As Simon Crean, the Minister for Primary Industries and Energy noted, the purpose of shifting the Farm Household Support provision to a separate piece of legislation was so that it would 'not undermine the achievement of our goal of a productive and sustainable farm sector' (Commonwealth of Australia, 1992c:2417). Due to low participation by farmers, the Farm Household Support Scheme was eventually scrapped in the 1996/97 Federal budget by a newly-elected conservative Liberal/National Coalition government, and replaced with the broader Farm Family Restart Scheme – discussed in a later section of this chapter.

The separation of 'welfare' and 'business' measures shows a realignment of rationalities and technologies of governing. Rather than encouraging all those farmers defined as marginal to improve their productivity, RAS was now to be aimed at those who constructed their farms as business enterprizes; an objective consistent with the recommendations of the Synapse Review. In other words, as Gow and Stayner (1995) note, its focus was beginning to be moved away from those on the margins of viability, and towards producers deemed capable of improving substantially their productivity through formal business techniques (see also Malcolm et al., 2000). From the perspective of an analytics of government, the key word here is *capability*, with authorities arguing that farmers required formal managerial skills in order for them to act in a more enterprizing and productive manner. Those farmers deemed to not have the necessary qualities were thereby constituted as unviable and unable to read market signals correctly (see Lawrence, 1992). Nevertheless, the separation of business and welfare measures was believed to improve not only the effectiveness of the RAS, but to result in less confusion about its overall objectives (see Hyde, 1997).

The above issues were raised in my interviews with Commonwealth public servants, who noted that a 'welfare' element in the RAS diminished its effectiveness and sent confused economic signals to farmers. For example:

> There has always been a welfare element to…[the RAS]…and that created a high degree of confusion in that it sent the completely wrong signal to farmers (Male, Branch Director level [2], DPIE).

> The efforts of policy-makers to try and keep the scheme confined to an efficiency-type scheme for productivity growth and so forth, has been muddied by two things. One is not separating out the welfare component…and the other one is not separating out sufficiently the carry-on type assistance which can be seen as helping farmers to get through a bad patch and not doing much else. (Male, Branch Director level [3], DPIE)

Interviewees at the Commonwealth level thus share an advanced liberal governmental rationality in which welfare and business measures are constituted as contradictory objectives (Rose, 1993, 1999; Dean, 1999). Forms of governing that seek to govern 'the social' are discredited as inhibiting efficiency and regarded as leading to expectations by farmers that governments will provide assistance in difficult situations. However, this does not mean that an advanced liberal form of governing dominates how the management of agricultural change is reflected upon as a problem. Chapter Five, for instance, seeks to show that social rationalities of governance continue to be significant in defining the limits of managed risk. The focus on individual management was again emphasized in an important review of farm adjustment in 1994.

Senate Report on Rural Adjustment, Rural Debt and Rural Reconstruction (1994)

On the 24th of March 1994 the Senate Rural and Regional Affairs and Transport References Committee (Commonwealth of Australia, 1994) was commissioned to inquire into the adequacy of farm adjustment and reconstruction programs, and the extent of rural debt in Australia. The Committee's findings on farm adjustment – specifically the adequacy of RAS

1993 – are of most relevance to this chapter and will therefore be given the greatest attention. Each of the components of RAS 1993 will now be examined, with the Committee's findings on each presented.

Farm productivity enhancement

The farm productivity enhancement measures of the RAS provided interest subsidies of up to 50 per cent of the cost of commercial finance for productivity related improvements such as debt restructuring and capital restructuring. These were first introduced in 1985 and regarded at the time as an efficient means of making adjustment assistance transparent, and ensuring that the financial benefits were passed on to farmers (Commonwealth of Australia, 1985b:1868). This was no longer the case by 1994, with a number of witnesses at the Senate Committee hearings expressing doubts concerning the effectiveness of interest subsidies as a means of improving farm viability, due to their targeting toward farmers with high levels of debt. For instance, the Australian Bureau of Agricultural and Resource Economics (ABARE) noted that 'interest subsidies distort adjustment processes and...it may be more efficient to pay subsidies in the form of a lump sum payment which would enable farms to allocate funds to the most appropriate need' (cited in Commonwealth of Australia, 1994:39). Applied specifically to drought, Gow (1996) elsewhere argues that interest subsidies for drought exceptional circumstances encourage 'perverse behaviour' by farmers, inhibiting more prudent land management. Hyde (1997) agrees with this point suggesting that interest subsidies distort the price of credit and encourage farmers under financial stress to make excessively risky decisions.

A number of the Commonwealth public servants interviewed as part of the research for this book viewed interest subsidies also as distortionary on farmers' management practices. For example:

> If you are going to help farmers you don't help them by using policy instruments that undermine the marketplace. Now interest subsidies are a perfect example of this. Interest subsidies encourage farmers to get something for nothing, encourage banks to be lazy in terms of their lending policies.... So you know you don't use interest subsidies, they just send the wrong signals altogether. If you want to provide support, it seems to me that you are far better providing direct grants for which farmers have to account for the dough. (Male, Branch Director level [2], DPIE)

> A lot of the problems of individual farmers in Queensland are because they have taken decisions on the basis that governments are allowing, and actually providing assistance through interest rate subsidies. The back of my mind says that you shouldn't do that because you are doing the wrong thing by the people that you are supposed to be helping because you are not allowing them to make a reasonable business decision. (Male, Senior Officer level, DPIE)

Views such as these were reflected in the Committee's recommendation that the Department of Primary Industries and Energy 'reassess whether interest subsidies are an effective instrument to facilitate rural adjustment' (Commonwealth of Australia, 1994:40). While the utility of interest subsidies was first questioned in 1992 by the Synapse Consulting Review of the Rural Adjustment Scheme, evidence to the Senate Committee emphasized this point and suggested they did little to smooth farm adjustment. From the perspective of an analytics of government, the questioning of interest rate subsidies represents part of the 'congenitally failing' nature of governing (Miller and Rose, 1990:10), in which programs rarely achieve their desired effects. Interest subsidies are examined further in a later section of this chapter.

Skills enhancement

The skill enhancement component of the RAS provided grants to assist farmers to improve their managerial skills. All evidence submitted to the Committee supported the views expressed in the Synapse Review, and the objectives of the 1993 RAS, that ongoing education and training was essential to improve farm management, and therefore productivity. The possession of a formal business plan was emphasized particularly as a crucial part of effective farm management, although the Committee stopped short of recommending that these should be mandatory in order to receive RAS assistance. The Committee recommended that:

- sound financial and management skills are the key to a robust, competitive, profitable and flexible farm sector. Accordingly, the Committee recommends that, in the medium to long term, skills enhancement, including education, training and access to professional advice, should become the principal component of the rural adjustment scheme;
- the Commonwealth and states develop procedures to ensure that recipients of RAS productivity and enhancement measures have sustainable business and farm plans (Commonwealth of Australia, 1994:42).

The perspectives from interviewees on the issues of farmer education, training and the possession of business plans will be discussed in the next section on the Land Management Task Force's evaluation of the National Property Management Planning Campaign.

Re-establishment grants

The re-establishment grant was designed originally to facilitate an 'orderly' exit for farmers from the industry. However, evidence collected by the Committee suggested that there were only 2,123 approvals for such grants between 1977 and 1994 (Commonwealth of Australia, 1994:Appendix 3). This is a very small proportion of approvals compared to other sections within the Scheme. Based upon evidence concerning the effectiveness of re-establishment grants, the Committee recommended that the level of the grant should be increased above $45,000 to facilitate a faster level of adjustment, or a modification of the assets threshold implemented so that farm families could exit the sector with more assets. Equally, the Committee recommended that a specific re-training component needed to be included as part of the re-establishment grants.

Exceptional circumstances

The purpose of exceptional circumstances provisions in the Rural Adjustment Scheme was to provide targeted assistance for circumstances that farmers could not reasonably be expected to plan for through normal risk management strategies. While the Committee noted a number of problems with the eligibility for, and lack of definition of, exceptional circumstances, it recommended that the provisions should continue as 'there are limits to risk management and self-reliance, and…there are circumstances for which even a prudent farmer cannot plan' (Commonwealth of Australia, 1994:52). The proposition that strategies of risk management have limits in their effectiveness will be explored in the next chapter. In support of exceptional circumstances provisions, Gary Goucher, the then Director of Policy at the National Farmers' Federation (NFF), noted:

> I can mount a very good argument, I think, that says [exceptional circumstances] should remain there, because commodity prices can be just as capricious as the weather, and mostly not just because of the natural forces of supply and demand. International commodity prices are frequently and substantially affected by what treasuries in other countries do rather than by the forces of supply and demand. (cited in Commonwealth of Australia, 1994:47)

Such an argument demonstrates a marked change from the traditionally 'dry' position of the National Farmers' Federation (Lawrence, 1987). Essentially it contests the advanced liberal rationalities of governing noted in previous sections of this chapter in which farmers were argued to receive the clearest signals to make economically rational decisions, only with the removal of distortionary (social) forms of assistance.

Despite recommendations that exceptional circumstances provisions should not be abandoned, evidence from a number of State governments argued that it was incompatible with RAS objectives, and should therefore be removed from the Scheme. In contrast, the Queensland government felt that the inclusion of exceptional circumstances in the Rural Adjustment Scheme was important in maintaining productivity and addressing drought. Nevertheless, the Committee, in its final recommendations, argued for the removal of exceptional circumstances from the Rural Adjustment Scheme due to its perceived inconsistency with RAS objectives, distorting its focus on productivity, profitability and sustainability (Commonwealth of Australia, 1994:52).

Political events during late 1994, however, brought to a halt any ambitions to remove exceptional circumstances from the RAS. Following a visit by the Prime Minister, Paul Keating, in that year to areas of eastern Australia that had been enduring an unusually extended period of dry conditions, a Drought Assistance Package (DAP) was released. This package incorporated a range of both welfare and business measures, including the popular Drought Relief Payment (DRP). At the time it was released, this package seemed to directly challenge the rationality of drought as a manageable risk. As one Commonwealth public servant argued, the introduction of the Drought Relief Payment:

> ...was an appropriate welfare response to the plight of farm families. You know, they were seen to be going into the worst drought for decades, and Keating said, 'look, you mean a lot to us, we want to keep you there, therefore we'll give you support for social reasons or welfare reasons' (Male, Branch Director level [3], DPIE).

In fact, the widespread support for this policy decision in both metropolitan and rural areas meant that any direct moves to challenge the 'exceptional' status of drought would have proved to be politically unpopular.

While the implementation of these measures is argued by some to have signalled the death knell of an economic rationalist approach to rural policy (Gow, 1997), it in fact provided further justification for economic authorities to argue for an improvement in farmers' managerial skills, and a more planned and long-term approach to drought. The establishment by the Prime Minister of a Land Management Task Force in November 1994, just two months after the release of the drought package, demonstrates that the personal capacities of farmers were seen as still the most appropriate site through which to address the problem of farm management.

Report of the Land Management Task Force (1995)

As noted above, the Land Management Task Force was established in November 1994 by the Labor Prime Minister, Paul Keating, in response to the ongoing drought throughout much of Queensland and New South Wales. It was set up to investigate primarily how the adoption of formal property management planning by agricultural producers could assist in planning for drought, and in improving sustainability and productivity over the longer-term (Commonwealth of Australia, 1995:117). The Property Management Planning Campaign (PMP) was launched originally by the Federal government in August 1992 as a joint Commonwealth-State initiative, and funded through the National Landcare Program. It represents a very specific advanced liberal technology of governing in that it sought to enhance the individual managerial abilities of producers so that they would have the capacities to deal with their financial and environmental risk in an increasingly deregulated economic environment (Lockie, 1999). Of interest here is how the Task Force, that reviewed this Program, sought to link property management planning, as a technology of risk management, with productivity, farm viability and sustainability.

One of the key arguments of the Land Management Task Force was that the adoption of formal business plans was crucial in responding to change, and in establishing a standardized code for industry best practice where producers could evaluate their management practices in a more rational manner. In other words, it was assumed that producers would, by employing formal property planning, know whether they were managing their properties in an efficient, sustainable and long-term viable manner. Accordingly, the Task Force defined property management planning as:

> An ongoing process for the total management of a farm business which assists producers to improve their profitability and achieve more sustainable resource use. It is regarded as a whole systems process whereby producers identify their personal objectives in the context of broader community aspirations. Then, by applying skills in business management, quality assurance, risk management, natural resource management including nature conservation, financial planning and control, marketing management, agricultural technology management and personal and staff management, they develop plans to fulfil those objectives. (Commonwealth of Australia, 1995: 17)

The adoption of these managerial skills was thought to be best achieved through producer participation in a series of voluntary introductory workshops focusing on business goal setting, natural resource evaluation and management, staff management, business planning, quality assurance, marketing and risk management. This was to be supplemented by ongoing education and training to improve farmers' decision-making capacities. According to the Task Force, the focus of the planning process 'should be on creating self reliant farm enterprises' (Commonwealth of Australia, 1995:102). The claimed role of government was to ensure not only that the broader economic and institutional environment was conducive to 'sound decision-making', but also that farmers were equipped with the 'tools, skills and knowledge...that will enhance the overall performance of Australian agriculture' (Commonwealth of Australia, 1995:102-103). Property management planning represented therefore a 'technology of agency' (Dean, 1997, 1999) through which management expertise attempted to act on farmers' calculative capacities so that they would evaluate their practices in an economically calculable manner.

The role of formal managerial skills in shaping the conduct of farmers in increasingly economically rational ways was emphasized also by public servants at the Commonwealth level, as well as the National Farmers' Federation. For instance:

> ...we have got a fairly large group of people out there who are not used to taking responsibility for themselves, not full responsibility, whose managerial skills are relatively weak, whose marketing skills are basically non-existent in cases, and I guess there is a case for time-limited support over a period, say a decade. And the kind of support I am talking about here is improving managerial skills, improving financial skills, improving their marketing skills and improving their environmental management skills. (Male, Branch Director level [2], DPIE)

> ...governments...have to try and provide a better base for farmers to be able to have the wherewithal to manage, and if they have the capacity to rely on their own resources, you have a much stronger sector with people who know how to manage risk and know they have to face change and deal with globalized markets. If you get into a debate with farmers they'll say that's just impossible. Whether that in the long-term leads to a secure future for farmers is debatable without attention to their management, production and market. (Male, Middle Management level, RIRDC)

> ...basically the farm business should either thrive or survive or go backwards according to how it's managed. What's been driving part of that process [self-reliance] of course is large increases in productivity in the agricultural sector, and increases in the price of capital equipment etcetera, which means that farms have to keep on getting bigger and be better managed in order to survive. (Male, Executive level, NFF)

However, the emphasis on managerial skills was by no means agreed upon by all, with some suggesting that education and training constitute only one aspect of 'good' farm management. For example:

> It is a very simplistic view that management skills are right. Yes, there are some people...doing very well who are good managers. There is also those doing okay, thanks very much, in terms of their income, but aren't particularly good managers. (Male, Senior Officer level, DPIE)

> ...farmers certainly do need to improve their financial management skills, there is no doubt about that. But I think that only touches the edge of the problem. I think there is a lot more to it than that. I don't think being smarter is necessarily going to ensure that people aren't going to be in problems in the future. (Male, Middle Management, QRAA)

> There is a pretty strong belief...that better business skills do make better farmers. That has been challenged too, I mean there is some pretty good primary producers that have no level of education at all, but they are certainly switched on to what is happening around the place. They just don't have formal education. Nonetheless, there is that thrust towards education like there has been right over Australia in all aspects of business and industry,

> education has been the buzzword. (Male, Senior Officer level, Qld. Office of Rural Communities)

The above statements draw attention to the contested nature of managerial skills. Thus, while the Rural Adjustment Scheme, drought policy and the National Property Management Planning Program, attempt to encourage formal education and training as the solution to the problems of farmers, these programs have effectively reconstituted risk as an individual responsibility to be managed in order for viability to be achieved. The link between farm management skills and productivity will be discussed in a later section of this chapter in the context of the Agriculture-Advancing Australia Policy.

Report of the Drought Policy Review Task Force (1996)

The managerial capacities of farmers were again placed under the spotlight in August 1995 when a Task Force was commissioned to review the response to drought of the Federal and State governments. This was a time when many rural producers were continuing to suffer from the effects of ongoing dry weather. In fact, numerous Local Government Areas – particularly in Queensland and New South Wales – remained drought-declared until well into 1998. The final Task Force Report was released in late 1996 following the election of a conservative Liberal/National Coalition government earlier that year. The Drought Policy Review Task Force, comprising Commonwealth and State officials, was commissioned to:

- examine existing drought response measures and evaluate the impact and appropriateness of these measures and the relative contributions of the Commonwealth and drought affected States, in meeting the objectives of the National Drought Policy; and,
- identify policy options for future droughts which would continue the development of a self-reliant, competitive and environmentally sustainable farming sector (Commonwealth of Australia, 1996:1).

These objectives make it clear that there was a continuing emphasis on enhancing the self-regulatory (economic) capacities of farmers, with the Task Force expected to evaluate policy responses to drought in terms of their effectiveness in achieving self-reliance.

Overall, the Task Force argued that the objectives of the 1992 National Drought Policy were of continuing importance in encouraging not only self-reliant approaches to drought management, but also the protection of environmental resources. In this respect, drought as a normal element of risk management was re-asserted, with the Task Force noting that there was 'no strong argument for government relief assistance to farming businesses in time of drought' (Commonwealth of Australia, 1996:9). However, the Task Force argued for three additional considerations to past government responses. The first concerned the apparent lack of understanding of drought, and of management strategies to deal with drought. An improved communications strategy was suggested in order to make farmers more aware of the principles of the National Drought Policy. Second, the Task Force believed that more sustained attention needed to be given to the welfare needs of drought-affected families and communities; a point recognised in the 1994 Drought Assistance Package. Nevertheless, this was seen as needing to be structured in such a way so as not to obscure the claimed need for exit on the part of low-income farmers. Third, the Task Force stated that drought policy measures that 'obstruct or slow the inevitable process of structural adjustment' should be avoided (Commonwealth of Australia, 1996:7). On this basis, the Task Force suggested that an integrated policy should be introduced for future droughts, taking into account farm business measures, environmental measures and welfare measures.

Farm Business Measures

The Task Force, in its report, endorsed strongly self-reliance. Transaction-based subsidies – such as fodder subsidies – and interest rate subsidies for debt were criticised as inconsistent with this objective, retarding farmers' incentives to take a more planned approach to drought, and impeding the claimed normal processes of structural change. On the other hand, measures for increasing farmers' capacities to manage risk were promoted strongly. Education and training, particularly through Property Management Planning (PMP) processes, were seen as essential in facilitating the attitudinal change among farmers necessary for a planned approach to risk management, and therefore improved viability. Once armed with business skills the outcome was argued to be 'improved farm management for long-term profitable farmers, and better off-farm opportunities for those exiting the farming sector' (Commonwealth of Australia, 1996:14). Farm financial planning was considered also a central element of drought policy. The Task Force recommended that the Income Equalization Deposits (IEDs) scheme, as a risk management tool, be simplified and made more commercialized so that uptake might be improved.

Environmental Measures

The removal of fodder subsidies and interest subsidies was seen as crucial by the Task Force in maintaining sustainable production activity on farms during periods of drought when natural resources were most vulnerable to degradation. Moreover, a more sustainable approach to drought was believed to be achievable through training, education, planning, research and development, and early warning systems.

Welfare Measures

While supporting welfare measures to assist low-income farm families, the Task Force emphasized that these should be provided on a 'needs only' basis. A slow rate of adjustment, exacerbated by the drought, was regarded as not in the interests of either farmers or the economy. The Task Force recommended, therefore, that as conditions improved, farmers should have access to an interim welfare payment called the Farm Recovery Scheme, that would be linked to adjustment out of farming. However, it was emphasised that this payment should ensure that farmers would clearly 'know' that they had a limited future in the industry, thus encouraging the adjustment process (Commonwealth of Australia, 1996:26).

The above arguments – all of which were later endorsed by Commonwealth and State Ministers – show how self-reliance was continuing to be constituted as a strategy for managing, in a enterprizing manner, both environmental and financial risk. Rather than providing relief after the event, the role of government was constructed as one of *facilitator*, ensuring that farmers had the necessary business skills to prepare for drought, deal with risk individually, and thus run a viable operation. Property management planning was here an essential governmental technology that provided the technical means of calculation to indicate whether farmers were managing their properties in the most economically viable manner. Equally, those producers who could not cope in this new environment, and who requested welfare assistance, were constituted as 'targeted populations' (Dean, 1997), argued by authorities to require clear signals so that they would be able to know that their practices were unsustainable in the long-term, and therefore take steps to leave the industry.

The issue of environmental sustainability was promoted in the Task Force Report as one of maintaining productive capacity of the property in the long-term. In this way, farm problems were constituted as a problem of *economic* sustainability with the prudent management of natural resources in the short-term, using formal property planning techniques, seen as necessary in order to offset long-term productivity losses. Those farmers who might have been unable to manage their property in this manner in the short-term were constituted as poor managers of risk and deemed unlikely to have a profitable future in the sector. Thus, the Task Force Report encouraged education, training and financial planning so that farmers could make the most 'rational' economic choices over whether they should leave or remain on their land. Such an assumption, however, neglects farmers' strong cultural attachment to farming as a lifestyle (Gow and Stayner, 1995), an issue that will be revisited in the next chapter.

The National Rural Finance Summit: *Activating Committee Report* (1996)

Self-reliance continued during 1996 to be deployed as a strategic means of differentiating between those producers deemed to have the capacities to manage risk, and those argued to have little future in the industry. A National Rural Finance Summit, held from July 3-5 sought to further categorize farmers in this way. Its main purpose was to discuss financial issues, 'with a view to the identification of strategies contributing to a long-term sustainability, profitability and international competitiveness of the farm sector' (Dept. of Primary Industries and Energy, 1998a:1). The Summit was held just four months after the March Federal election in which the conservative Liberal/National Party Coalition, in Opposition since 1982, won a landslide victory.

The new government's approach to the governing of agricultural change was much more business-focused, with an increased emphasis on entrepreneurialism, prudent financial management and fiscal responsibility. Thus, the general themes of the National Rural Finance Summit included structural adjustment, farm business management, debt servicing, financial servicing, skills and education, taxation, welfare and infrastructure. There were 135 recommendations developed as a result of the Summit. Rather than examine the many outcomes of this conference in detail, it is of more use to the purposes of this chapter to examine the findings of the Summit Activating Committee – a body established to provide 'actionable advice' to the Minister for Primary Industries and Energy on the recommendations resulting from the Summit.

An Activating Committee was established following the National Rural Finance Summit. The objectives of this committee were to:

- review and assess Summit recommendations and other issues for consideration;
- advise Ministers of recommendations that might be incorporated immediately into government policies and programs;

- further analyze and develop for the Government's consideration, proposals requiring changes to legislation and/or formal incorporation into the Government's rural policy framework;
- liaise with stakeholders, including the National Farmers' Federation and the Australian Bankers' Association, in relation to those recommendations that fall within the roles and responsibilities of the respective stakeholders; and,
- contribute to the development of a Business Plan for Australian Agriculture (Dept. of Primary Industries and Energy, 1998a:2).

Of most relevance for this chapter is the Activating Committee's recommendations on education and training, and adjustment issues, contained in the final report.

The Committee argued, following recommendations made at the Summit, that training needed to be given most priority in the areas of financial management, marketing, natural resource management, risk management and succession planning. They recommended that government and industry should direct more funding towards personal, group and community activities that promoted a 'culture of change', integrated family farm business planning, comparative analysis and benchmarking, and improvements in communication and leadership skills (Dept. of Primary Industries and Energy, 1998b:2). Also highlighted was the need for education and training to be 'industry driven', farmers' skills and knowledge recognized through participatory activities to enhance local expertise and training abilities, flexible delivery mechanisms suited to rural circumstances, and government support of high-quality education and training in rural areas (Dept. of Primary Industries and Energy, 1998b:3).

Of further significance were the Committee's findings on farm adjustment. The Committee believed that independent advice and farm business planning were essential in improving farmers' ability to manage change, increase farm viability and, for those farm families in severe financial circumstances, make the decision early to leave. On this basis it recommended an adjustment package to provide:

- independent financial advice, to determine the viability of the farm business and to assess options;
- for those farmers in severe financial circumstances, access to a payment that is income and non-farm asset means tested, for a two year transitional period, subject to entering an agreement to either, 1) Stay on the farm and receive government assistance to further the farmer's education and training, improve farm business planning and undertake other activities aimed at increasing farm viability; or 2) Sell the farm and receive career counselling and retraining, and a relocation grant of up to $45,000 payable to farmers with a net asset base of up to $90,000 (Dept. of Primary Industries and Energy, 1998c:2-3).

The Committee further recommended that the Rural Adjustment Scheme should be re-focused with a greater emphasis on enhancing the ability of farmers to manage change by providing support for farm business planning, increasing skills and knowledge, and promotion of cultural and attitudinal change.

The Activating Committee's recommendations had a clear focus on enhancing the self-regulatory capacities of farmers so that they could assess their viability according to techniques of formal business planning. While it is claimed that change should be industry driven, this is the very type of rationality upon which advanced liberal forms of governing are based, where optimization of economic life is pursued through enhancing the individual 'choices' of autonomous actors. One becomes 'empowered' to make these choices through techniques of economic calculation (Rose, 1993, 1999). This is certainly the case here where the role of state agencies is constructed as promoting cultural change so that producers 'know' the most (economically) efficient option to take in managing their properties.

The recommendations of the Rural Finance Summit Activating Committee formed the basis of a proposal by the Primary Industries Minister, John Anderson, for an Action Plan for Australian Agriculture. Development of this plan is beyond the time frame covered by this book. More significant for the purposes of this chapter, however, are the ways in which the Committee's recommendations had an impact on the types of issues seen as relevant in the Mid-Term Review of the 1993 Rural Adjustment Scheme.

Mid-Term Review of the 1993 Rural Adjustment Scheme (1996-1997)

The Rural Adjustment Act of 1992 contained provisions for a mid-term review of the RAS in 1996. In keeping with these provisions, a review was commenced in November 1996, with the final report released in May 1997. The review of the RAS took place at a time of continuing drought in eastern Australia and unfavourable terms-of-trade for many farmers. Election of the conservative Liberal/National Coalition government in March 1996 had placed microeconomic reform and greater competition – through the National Competition Policy – as central priorities on the political agenda, thereby structuring the discursive limits of debate over future options for the agricultural sector. Business management by farmers was essentially the key issue shaping how the Review Committee was asked to evaluate the effectiveness of the Rural Adjustment Scheme.

The Review Committee's terms of reference were to 'report to the Minister for Primary Industries and Energy on the future adjustment challenges facing Australia's agricultural sector, the role of risk management in on-farm business strategies and the role, if any, of government intervention' (McColl et al., 1997:145). These terms of reference were to take into account the appropriateness, performance, management, administration and delivery of the Rural Adjustment Scheme. In pursuing the terms of reference, the Review Committee were asked to consider the Report of the Senate Rural and Regional Affairs and Transport Committee on Rural Adjustment, Rural Debt and Rural Reconstruction (Commonwealth of Australia, 1994); the Drought Policy Review (Commonwealth of Australia, 1996); and the outcomes of the National Rural Finance Summit (Dept. of Primary Industries and Energy, 1998a, 1998b, 1998c). The Review Committee's final recommendations were, overall, consistent with these earlier reports, and argued for a restructuring of adjustment assistance consistent with the RAS objectives of risk management. These criticisms of the RAS, and the Committee's key recommendations, are outlined below.

The most significant recommendation by the Committee was that interest rate subsidies for productivity improvement and for exceptional circumstances, estimated to be up to $40 million a year (see McKenzie, 1997a), should be abolished. Evidence collected by the Committee suggested that exceptional circumstances interest subsidies, in particular, were ineffective, divisive and encouraged production during periods when it could least be sustained. It was claimed also that such subsidies lessened the incentive for self-reliance as they created an expectation that governments would provide support in adverse conditions (McColl et al., 1997:112). In fact, the Committee argued that interest subsidies might have impeded claimed natural adjustment processes by preventing the implementation of a more prepared approach to farm management. This was a view consistent with the Senate Report (Commonwealth of Australia, 1994), the Drought Policy Review Task Force (Commonwealth of Australia, 1996) and the perspectives of many of those interviewed as part of the research for this book. Interest subsidies for productivity improvement were subsequently scrapped in the 1997 Federal budget. However, they were retained for Drought Exceptional Circumstances.

The abolition of interest subsidies occurred despite their important role in allowing farmers to restructure their debt during periods of drought and low commodity prices (McKenzie, 1997b). Nevertheless, the Review Committee sought to question such subsidies as based upon outmoded protectionist principles that created dependency and impeded the establishment of a more entrepreneurial and individualistic relation to risk. In this way, advanced liberal technologies that sought to foster more enterprizing attitudes were constituted as the only 'truthful' means for farmers to manage their properties.

Drawing on the recommendations of previous reports and reviews, there was a firm focus on the role of education and training in fostering improved farm management. The Review claimed that once the decision-making capacity of farmers was strengthened, they could more effectively deal with the management of risk and adjustment pressures. Skills enhancement provisions in the RAS were argued to have too little emphasis, receiving only a very small proportion of the Scheme's overall funding. The Review Committee recommended that a more central focus on enhancing skills would improve farmer's managerial capacities and thus their productivity. Formal education was seen to enable farmers to 'be self-reliant in managing their business or, for some, to recognise the inevitability of leaving agriculture and seeking off-farm pursuits and income' (McColl et al., 1997:85). Thus, the professional advice of economic expertise was believed to enhance farmers' ability to assess their future viability in the industry and enable the most rational decisions to be made. From this perspective, economic expertise was to have a central authoritative role in the transformation of farmers into active managers of risk. Moreover, the Committee argued that support for farm management should be better integrated with property management planning.

The Review Committee recommended that RAS 1993 should be terminated and replaced with a new programme 'that addresses the adjustment issues of management skills, farmer re-establishment, savings and welfare' (McColl et al., 1997:119). This was to consist of:

- A Farm Business Improvement Scheme (FarmBIS), the purpose of which would be to provide grants to community and groups and farmers to promote a culture of continuous learning and enhance the capacity of farm managers to identify, acquire and apply the skills and information needed to improve the performance of their farm business, both in terms of profitability and sustainability.
- A Farm Re-establishment Scheme, constituting a package of business advice and assessment, and a $45,000 re-establishment grant available for two years, to encourage farmers who wished to leave farming to do so as early as possible.
- Development of improved instruments to promote farm savings.

- More responsive welfare arrangements. (derived from McColl et al., 1997:118-120).

As with previous reports and reviews, the Mid-Term RAS Review Committee endorsed a self-reliant approach to managing climatic and financial risk. Farmers' self-regulatory capacities were seen as needing to be enhanced through formal business skills and education, by which they would be able to assess their conduct in a more economically calculable framework. Thus, the Review Committee encouraged, in the same manner as previous reports and reviews, those technologies of governing that would enable the practices of farmers to be rendered visible, so that their agency could be shaped in a manner consistent with advanced liberal rationalities of managed risk.

Agriculture - Advancing Australia (1997)

The outcomes of the RAS Mid-Term Review, the National Rural Finance Summit Activating Committee, the Drought Policy Review Task Force, and the Report of the Senate Rural and Regional Affairs and Transport References Committee, were all significant in the formulation of the Commonwealth's *Agriculture - Advancing Australia* policy (or the 'Triple A' package) announced on 14 September 1997. While remaining committed to a strategy of self-reliance, this policy saw the abolition of the Rural Adjustment Scheme, with farm business objectives separated clearly from welfare measures. The separation of welfare and business measures into different programs was, generally, endorsed by respondents interviewed in the course of this study. For example:

> Well I think it probably had more of a welfare rationale than was stated. But that's tended to prevail all the way through, and it's probably been the cause of quite a lot of confusion later on where some of the states still run it partially as a welfare scheme and partially as a business assistance scheme. This led to confusion of objective and outcomes. It's not always in the best interests of the business if you're trying to achieve welfare objectives through assisting business, and that was an issue that the McColl Report [Review of RAS 1992] stressed. (Male, Middle Management level, RIRDC)

> I think we really need to move away from the idea that...[rural adjustment policy]...is a welfare-based system, to a system of management development and business skills development (Male, Executive Member [2], CUA).

The Triple A package was argued by the Federal government to promote 'a more competitive, sustainable and profitable' rural sector through four objectives:

- to help individual farm businesses profit from change;
- to ensure the farm sector has access to an adequate welfare safety net;
- to provide incentives for ongoing farm adjustment; and,
- to encourage social and economic development in rural areas (Dept. of Primary Industries and Energy, 1998d:3).

It is the first three objectives that are of most interest here as they are linked to measures available previously through the Rural Adjustment Scheme. Consistent with the recommendations of the RAS Mid-Term Review (McColl et al., 1997), the business elements that were implemented through the Triple A package included a Farm Management Deposits Scheme and a Farm Business Improvement Program (FarmBis). The welfare elements of the package included a Farm Family Restart Scheme, Exceptional Circumstances and a Retirement Assistance Scheme. Each of these programs will be now examined except the Retirement Assistance for Farmers Scheme, as the inter-generational succession issues that it raises are beyond the scope of this book.

Farm Management Deposit Scheme (FMD)

The FMD Scheme, which commenced in July 1998, was designed to replace the former Farm Management Bonds Scheme (FMB) and the Income Equalization Deposits Scheme (IEDs). It was established to provide a mechanism for enhancing farmers' abilities to manage financial risk, and was administered by financial institutions. While the transfer of administrative responsibility to financial institutions might appear to indicate that the FMD Scheme was a fully privatized form of financial risk management, this was not the case. Rather, the Commonwealth government saw its role as 'providing the tools to assist farmers plan for future uncertainties' (Dept. of Primary Industries and Energy, 1998d:4). In other words, while the Commonwealth was responsible for the policy instruments to be used within the scheme, the role of banks was to distribute the funds in accordance with Federal government guidelines. Thus, FMDs represented a technology of agency for state agencies to act on farmers' self-regulatory capacities 'at a distance' (Miller and Rose, 1990; Rose and Miller, 1992) so that they would evaluate their financial practices in seemingly more rational ways.

Farm Business Improvement Program (FarmBis)

The Farm Business Improvement Program (FarmBis) was designed to provide financial support for skills development, farm financial planning and advice, farm performance benchmarking, quality assurance, marketing, risk management, rural leadership development and natural resource management. These were claimed to be needed in order to improve the skill base, profitability and sustainability of farm businesses (Dept. of Primary Industries and Energy, 1998d). One of the assumptions underpinning this program was that formal training contributed to improved profitability and environmental sustainability. However, as argued earlier, in the context of the Synapse Consulting Review of the 1989 Rural Adjustment Scheme, this is a highly contested issue with the influence of formal risk management skills on farm financial performance unclear (see Hyde, 1997).

What is frequently overlooked in arguments that promote improved formal skills as the key to efficient and sustainable farms is that changes in farm management practice are influenced by a diversity of factors (see Kilpatrick, 1996; Bamberry, Dunn and Lamont, 1997; Garnaut and Lim-Applegate, 1998). For instance, according to Bamberry et al. (1997), programs that seek to link productivity with formal management practices need to take into account farmers' own understandings of their practices, particularly the role of non-vocational training, 'local' knowledge, and 'innate ability'. This means that the 'local' knowledges of producers need to be considered more closely in programs that attempt to govern through the agency of farmers. Nevertheless, in its present form, the promotion of formal managerial skills might be seen as an attempt to govern farm practice in an advanced liberal way (Rose, 1993). That is to say, formal training represents a technological means in which the use of economic management techniques is constituted as the most truthful way of assessing (and altering) a property's viability.

Farm Family Restart Scheme

The Farm Family Restart Scheme was aimed towards those low-income farmers experiencing financial hardship, or those who wished to leave the industry. Three programs were offered within the Scheme: First, Restart Income Support, that aimed to provide twelve months of income support while farmers decided whether or not to leave the industry; second, a re-establishment grant of up to $45,000 for those farmers wishing to leave the sector; and third, professional advice up to the value of $3,000 available to farmers wishing to explore future options outside the industry. The counselling component of this scheme is the most sociologically significant point here that has not yet been discussed from the perspective of an analytics of government.

As part of the Farm Family Restart Scheme, professional financial advice in one or more of the areas pertaining to financial planning, business management, career options, legal matters, and personal and family matters, was compulsory for those farmers on income support, or for those unable to sell their property within its first three months on the market. Such advice was intended to provide an assessment of the property's viability, and encourage the development of a formal plan for the future (Dept. of Primary Industries and Energy, 1998d). In making counselling compulsory, farmers who could not become self-reliant were effectively constituted as non-viable and seen as having neither the necessary skills nor entrepreneurial drive to remain in the industry. Financial incentives encouraged farmers to evaluate their managerial practices and options in economic terms and, in turn, formulate the seemingly most rational response to their circumstances. In this way, the counselling component of the Farm Family Restart Scheme represented a technological means through which farmers could come to know that their practices were non-viable in the long term. Their exit was thereby governed through individual choice rather than via more coercive forms of intervention.

Exceptional Circumstances

The continuation of exceptional circumstances provisions in the *Agriculture - Advancing Australia* package represented a recognition by government that there were some situations that no producer could be expected to reasonably plan for and manage. These provisions included: interest subsidies of up to 100 per cent; an Exceptional Circumstances Relief payment, subject to assets test; and, possible eligibility for additional social security support (Dept. of Primary Industries and Energy, 1998d). While programs of exceptional circumstances are regarded in official political documents as inconsistent with individual risk management, their focus on long-term viable producers does not make them fit neatly into the welfare category. Chapter Five analyses this issue in more detail and explores the significance of exceptional circumstances provisions in shaping the limits of individual risk management.

The creation of FarmBis to encourage enhanced formal management skills, and the establishment of a Farm Management Deposits (FMDs) Scheme to foster financial self-reliance, show the importance of advanced liberal rationalities and technologies of governing in constituting both the environment within which farmers made their decisions, and the types of conduct deemed rational within that environment. Equally, those farmers unable to take a more personalised approach to risk were constituted, through the Farm Family Restart

Scheme, as unviable and counselled to reassess their future in farming. This chapter now turns to the ways in which self-reliance has been shaped and reconstituted by State administrative bodies.

STATE ADMINISTRATION OF SELF-RELIANCE

As emphasized in previous chapters, state authorities play a crucial role in governing agricultural change in Australia. This is due to the fact that under Section 96 of the Constitution the Commonwealth may provide financial grants to any State, under conditions that the Parliament sees fit. Since authorities at a Commonwealth level of government define the managerial skills of some farmers as a problem of national economic significance, the co-operation of the States has been sought for the purposes of managing this issue. However, as the States have a long tradition of resisting attempts by the Commonwealth to overtly control their affairs, any agreements must be flexible, but in a way that achieves national objectives. Thus, while State authorities – who administer programs such as the RAS – believe that they should have flexibility in the interpretation and application of Federal guidelines, Commonwealth agencies want national objectives to be compromised as little as possible. This is a significant issue, as it has a direct bearing on whether the conduct of State authorities is consistent with Commonwealth objectives, and thus how self-reliance is governed.

According to the RAS 1992 guidelines the Commonwealth's role is to:

- define the overall purpose and objectives of the RAS.
- specify the outcomes expected from the application of funds.
- give policy directives to the States and Territories within which the RAS is to operate.
- establish a reporting mechanism to permit the monitoring and assessment of the efficiency and effectiveness of the operation of the RAS.

On the other hand the States, through relevant departments or agencies, must:

- properly manage funds provided under the RAS.
- assume responsibility and accountability for the achievement of the objectives of the RAS in accordance with the Commonwealth's policy guidelines. (derived from Rural Adjustment Scheme Advisory Council, 1996:19)

The above guidelines encourage a degree of flexibility, with the States able to decide which elements and forms of assistance within the RAS are the most appropriate to their requirements. As two of the Commonwealth public servants interviewed in the course of the research for this book noted:

> Well, we've given the states a fair sort of responsibility in terms of the interpretation of some of the guidelines, and there has been different interpretations between different states. So, you'll find inconsistencies between the States' approaches. And while the guidelines are probably loose enough for them to do that, in fact the guidelines have been reasonably loose and the states have used that to their advantage at times in terms of the

> way they've done things. It gets back to the situation of responsibility for the operation. I think the guidelines probably could have been tightened a bit. (Male, Branch Director level [1], DPIE)
>
> We have taken the line that we see the program [RAS] as being a national program delivered regionally, or delivered on a state basis, and that allows the states to be closer to the action to fine tune the program and work out any differences (Male, Assistant Secretary level, DPIE).

While endorsing the need for administrative flexibility at the State level, a Queensland public servant noted that it would be very difficult to achieve standardization across the different states:

> RAS was a bit of an education in how do you have a nation-wide policy that you administer in seven different jurisdictions with seven different organizations and how do you make that consistent? But certainly there were problems, and a lot of that was about the Commonwealth trying, and I could see what they were doing, they were saying, 'we have a lot of different conditions in terms of trying to implement a national policy, so we are trying to not be incredibly prescriptive but we are going to try and provide guidelines for how training grants are administered, or how productivity is administered, and state organizations can take those guidelines and say I think these guidelines mean this'. (Female, Senior Management level, QDPI)

Reviews of the RAS in 1992, 1994 and 1997 noted that while there were often differences in interpretation between the States, this was an inevitable part of taking into account the different regional adjustment requirements. Nevertheless, the Senate Review in 1994 (Commonwealth of Australia, 1994) and Mid-Term Review of the RAS in 1997 (McColl et al., 1997) noted that Commonwealth guidelines needed to be made more prescriptive to avoid anomalies and ambiguities in interpretation, and introduce a greater degree of standardization among the States. In this respect, the introduction in 1995 of an accounting and administrative procedures manual by the Department of Primary Industries and Energy represented a significant attempt to introduce a more 'transparent' and 'truthful' way of interpreting Commonwealth guidelines. This manual is an example of what Dean (1999) calls 'technologies of performance' that exercise scrutiny over the conduct of authority. Thus, for the agency of farmers to be shaped effectively in an advanced liberal way, attempts were made to transform the procedures of calculation, classification and accounting used by authorities into a stable, mobile comparable and combinable form (see Latour, 1988).

The assessment of a property's viability is here of relevance. While various reports, reviews and policy statements have sought to more 'precisely' define the boundaries of viability, according to formal procedures of business management, the Synapse Review of the RAS (1992), Senate Report (Commonwealth of Australia, 1994), and Mid-Term Review (McColl et al., 1997) all found a lack of objective and consistently-applied indicators for the assessment of RAS clients. Thus, attempts to introduce greater co-ordination among State administrative bodies appear to have encountered persistent problems. As one Queensland public servant noted, the complexity of issues such as how viability is defined makes it difficult to prescribe standardized procedures:

> People can say how do you assess viability? That is a grey area. It is on a case-by-case basis. No two farms are the same, no two farmers are the same. So you can't just simply have a rule for across-the-board to say that if your debt load exceeds X amount of dollars or if your debt load exceeds so much per breeder or so much per acre or whatever, you are non-viable. (Male, Middle Management level, QRAA)

It appears that attempts to introduce administrative and financial standards so that State bodies assess RAS applicants in more economically rational ways, consistent with the objective of self-reliance, have been shaped variously according to how understandings of farm viability are constituted at the State level. This does not mean that State administrative bodies have rejected self-reliance as a desirable objective. In fact, reports on, and reviews of, farm adjustment since 1990 show widespread endorsement by both conservative and reformist parties of self-reliance as a worthwhile, and indeed necessary, goal for Australian agriculture. However, its governance has been shaped by agri-political conditions in each State, the priorities of the bodies responsible for administration of the RAS, and the nature of specific regional problems. An in-depth analysis of how different States have sought to translate viability is beyond the scope of this book. Therefore, the following section of this chapter focuses primarily on the State of Queensland to investigate the ways in which Commonwealth objectives have been rendered knowable and governable to those authorities responsible for administering rural adjustment and drought policies.

Self-Reliance and the Queensland Government

Since 1992 the Queensland Government has supported the Commonwealth objective of self-reliance. The Queensland State government's support for a self-reliant approach to risk management emerged with the release of a policy paper in March 1992 entitled *Drought: Managing for Self Reliance* (Queensland Government, 1992). This publication, released when a Labor government was in office both Federally and in Queensland, was consistent with the National Drought Policy in that it endorsed generally a self-reliant approach to the management of risk by farmers, emphasizing the importance of financial and business planning in addressing environmental issues. As the policy paper stated:

> Self-reliance is inextricably linked to the concept of economic and environmental sustainability. It involves the adoption of pre-emptive strategies to minimise the social, economic and environmental impact of severe climatic events... The planning process should ensure that the productive capacity of the property can be maintained at an adequate level (without environmental degradation) during drought, or that financial reserves are sufficient to maintain the enterprise until conditions improve. Self-reliance clearly implies that in the longer term, Government assistance should not be necessary to sustain or support the enterprise during drought. (Queensland Government, 1992:9)

Farm sustainability was here linked clearly to productivity improvement. It states explicitly that pre-emptive business planning results in improved environmental management. To achieve the twin objectives of financial and ecological sustainability a number of strategies were suggested including: income smoothing, taxation incentives, the maintenance of the Rural Adjustment Scheme, a systems approach to property management, climate and commodity prediction services, and business education and training. It was assumed that these 'technologies of agency' enabled the capacities of farmers to be shaped so that they could properly manage risk. Nevertheless, it was recognized that even if total self-reliance were achievable, a role for the Queensland Government would remain necessary in maintaining support for those climatic events defined as severe. Hence, some forms of risk were constructed as beyond the normal managerial capacities of farmers. This issue is discussed in more detail in Chapter Five.

Despite the political objective of self-reliance, 66 per cent of total RAS approvals in Queensland from 1992/93 to 1995/96 were for drought exceptional circumstances (McColl et al., 1997), a provision seen by authorities as generally inconsistent with the rationalities and technologies of individual risk management. This raises the issue of how administrative bodies in Queensland have sought to constitute Commonwealth objectives of individual risk management. While supporting self-reliance as a desirable, and achievable objective, participants interviewed for this study at the Queensland State level of government, believed that the contingencies associated with drought had constrained the adoption of those education and training programs aimed towards enhancing the managerial capacities of producers. For example:

> The majority of our assistance in Queensland has gone to drought assistance, something like 150 million dollars in the last four years has been provided to producers who are struggling to overcome the drought... It [self-reliance] was a good move, but the timing of it couldn't have been worse. It is just unfortunate that it came that way. (Male, Middle Management level, QRAA)

> In Queensland...we have been more seriously affected by drought and so drought has really taken over everything in the last five years...[P]robably around two-thirds of all the RAS money in Australia has gone to Queensland in the last few years, and the most practical way of managing that funding is to put it into interest rate subsidies... If things were going really well in Queensland agriculturally, right across the board, there might be more scope to increase the focus on education (Male, Senior Officer level, Qld. Office of Rural Communities).

Most Commonwealth public servants interviewed believed that administration of the RAS in Queensland was, overall, consistent with Federal objectives. However, one interviewee believed that less emphasis should have been placed on interest subsidies:

> I think they [Queensland] would like to believe they were [consistent], but unfortunately they see that as the big problem, the amount of debt that's there... And they see that as the problem that needs to be addressed before they go anywhere else. We don't see that as the way to go. What little money we have we need to focus on profitability. And as I said before, Queensland is probably the most conservative of all states in that situation, and it seems to be a co-option that's shared by the government as well as the constituents. (Male, Senior Officer level, DPIE)

While Queensland, compared to the other States, is not necessarily abnormal in its approach to administering the RAS, there have been moves to adopt a more business-oriented focus consistent with Commonwealth objectives of self-reliance.

The creation in 1994 of the Queensland Rural Adjustment Authority (QRAA) can be seen as an attempt to introduce a more economically rational approach to the assessment of farmers applying for rural adjustment programs. Between 1990 and 1993 rural adjustment assistance was administered through the Queensland Industry Development Corporation's (QIDC) Government Schemes Division. The objective of the Corporation was to 'facilitate, encourage and promote the development and expansion of primary, secondary and tertiary industries in Queensland, with a view to enhancing economic growth and employment opportunities in the State' (Parliament of Queensland, 1986a:3456). Following corporatization of the QIDC in 1994, a new authority for delivering assistance was created that was not a commercial lender in its own right. QRAA was created in September 1994 under the auspices of the Queensland Labor Party. Its creation was supported by the conservative National and Liberal parties.

The key objective of QRAA was 'to deliver financial and other assistance primarily targeted at fostering the development of a more productive rural sector' (Parliament of Queensland, 1994a:s.3.(1)). The Authority's key functions were to:

- put approved assistance schemes into effect by ensuring that the schemes are properly and fairly administered, and giving the assistance for which the schemes provide;
- to arrange for education, training and advice to applicants under approved assistance schemes;
- to negotiate for assistance funding with all levels of government;
- to carry out research, and develop policies on, issues affecting persons likely to receive adjustment assistance; and,
- to review, and give advice to the Minister, on assistance schemes and other conditions in the rural sector (Parliament of Queensland, 1994a:s.8(1), 8(2)).

Its specific role was to deliver assistance under the Rural Adjustment Scheme, and under state programs such as the Primary Industry Productivity Enhancement Scheme (PIPES). This latter scheme falls under the Farm Productivity Enhancement objectives of the RAS and provides concessional finance to producers unable to borrow from normal commercial sources for such purposes as farm build-up, rationalization and capital works. The differences in objectives between the two authorities are not difficult to note. While the QIDC was aimed towards economic expansion and employment, QRAA had a clearly defined focus on productivity and business management; goals that make its activities consistent with the RAS 1993 objectives and an advanced liberal rationality of governing.

The shift in focus of the new Authority did not go unnoticed by many members during the State parliamentary debate over the establishment of the Queensland Rural Adjustment Authority. For instance, Mark Stoneman – National Party Member for Burdekin – demonstrating the agrarian fundamentalism for which the Party is known, stated that:

> ...the prime objective of any support must be to sustain production, and the basic unit of production in the primary sector, that is, the family farm. Without either of those units rural Australia ceases to exist. It is the means by which the nation was made, and it is how the nation will continue to be sustained regardless of the growth of industries such as tourism and mining. (Parliament of Queensland, 1994b:9248)

Equally, Howard Hobbs, National Member for Warrego, noted also that the encouragement of self-reliance would have serious implications for family farming. He claimed that the short-term assistance provided through QRAA was inadequate for rural industries, and that longer-term assistance was needed so that rural Australians could remain on their properties (Parliament of Queensland, 1994b:9270).

Since 1992, attempts have been made by various authorities to align State administrative procedures with Commonwealth objectives. However, a contradiction appears to have emerged between the Commonwealth's desire to standardize, and make 'transparent', assessment across the States, and the perceived threat of removing administrative flexibility. Applied to Queensland, such flexibility is valued, particularly in terms of ensuring that family farming as a rural way of life is maintained. While specific technologies of performance, such as accounting procedures, have been introduced in an attempt to achieve conformity, it seems that in the absence of direct intervention, individual States will continue to govern farm viability according to the variable political, economic and climatic conditions of specific regions. This also appears to be the case with agricultural producer groups to which this chapter now turns.

AGRICULTURAL PRODUCER GROUPS AND 'SELF-RELIANCE'

Agricultural producer groups in Australia have played a prominent role in promoting a self-reliant approach to the governing of agricultural change. Such groups have traditionally been split along interventionist/non-interventionist lines. While large woolgrowers and graziers have generally endorsed a non-interventionist and free trade approach to agricultural regulation, farmers have tended to support direct government action in their affairs (Encel, 1970; Hefford, 1985). One might expect these same divisions to exist among producer organizations concerning self-reliance. As noted earlier in this chapter, however, there appears to be a widespread endorsement (albeit in principle) by many producer organizations that self-reliance is a desirable and necessary goal if Australian agriculture is to prosper. This is not entirely surprising given the prominent place of independence and hard work among farmers, and the claimed significance of agriculture to national growth (see Aitkin, 1988).

Nevertheless, while notions of individual enterprize and 'being one's own boss' appear to sit well with the individualism of advanced liberal rationalities of governing (see Lockie, 1999), there is an historical tension between the desire by producers to manage independently, and the claimed right to government assistance (Marshall, 1985). Of relevance then is how producer groups have sought to constitute self-reliance and its relationship to advanced liberal rationalities of governing. The views of the peak Australian farming body, the National Farmers' Federation (NFF) will first be examined followed by the perspectives of participants interviewed in the course of this study.

National Farmers' Federation: New Horizons

The National Farmers' Federation (NFF) views on the governance of agricultural change were outlined in a key statement of policy in 1993, entitled *New Horizons* (National Farmers' Federation, 1993). This document endorsed adjustment as an inevitable process to improve agricultural efficiency and productivity. Government intervention, however, was supported as an integral part of this process to recognize 'that imperfections in capital and product markets and intervention elsewhere by government can produce periods in which autonomous adjustment is inadequate' (National Farmers' Federation, 1993:122). For the NFF, adjustment included changes not only in the structure of the farming sector's physical resources, but also in human skills. Individual risk management here assumed significance in constituting the managerial capacities of farmers as the legitimate sites to facilitate adjustment. As the NFF (1993:122) noted, 'The increasing complexity of marketing, financing, management and technology in agriculture demand on-going development of the operator's skills in order to maintain productivity growth'.

The enhancing of producers' self-regulatory capacities, for the purposes of 'effective' risk management, was here consistent with the advanced liberal technologies of agency promoted in Commonwealth government documents. Nevertheless, the NFF recognized that where self-reliance was to be a policy objective, clear provision should be made for those circumstances – such as droughts defined as severe – beyond the planning capacities of producers (National Farmers' Federation, 1993:123). In this way, the NFF were effectively defining the limits of self-reliance and the terrain of governing in which technologies of risk management were deemed problematic. Finally, the NFF supported the removal of Household Support from the Rural Adjustment Scheme in 1992. Household support was claimed to have led to a misunderstanding and confusion over the Scheme's objectives of productivity and competitiveness.

While the National Farmers' Federation endorsed increased productivity and a self-reliant approach to farm management, in the interests of competition and freer trade, State-based organizations adopt a slightly different viewpoint that is indicative of the tension noted above between individual enterprize and a desire for government assistance (see Marshall, 1985). A number of public servants interviewed noted this tension, particularly among Queensland producer groups. For example:

> NFF have [accepted self-reliance], the state ones haven't. The New South Wales Farmers' Federation has not, they are still looking for transaction-based subsidies, so they still want assistance to shift stock from one point to another, to bring in fodder etcetera. Those type of transaction-based subsidies are anathema to the direction that the Commonwealth's going. So while I don't believe that they have taken it on fully, they use the rhetoric but they do not use the actions. (Male, Branch Director level [1], DPIE)

> ...the farm organizations in particular like the NFF have been pretty strongly supportive of a stand-on-their-own-two-feet, we'll run this as a business thank you very much, in terms of day-to-day. But when it's come to exceptional circumstances, in other words the fluctuations in the day-to-day environment, they've been pretty supportive of exceptional circumstances-type support. So there's a philosophical tension there, you know, the more exceptional circumstances support you provide, the less you're able to stand on your own... Queensland's a classic example of that. (Male, Branch Director level [3], DPIE)

> ...you'd find farm organizations that would, most of them would probably say, 'yes, that is the direction we have to move' [towards self-reliance]. But my impression is that a lot of them haven't been particularly pro-active about it. I think the NFF is starting to be a bit more now, but generally they have tended to follow the lead. Quite often you'll find that they will support the status quo that is really a policy that they thought at the last round

> would work... I think most of them would generally agree that it's the centre of their objectives, but they would disagree over what governments can effectively do in the market. (Male, Middle Management level, RIRDC)

The contradictions noted in the above statements appear to be borne out by representatives interviewed from Queensland producer groups. For example:

> We have always taken the battle up to both State and Federal governments. In that regard, we have always maintained that it is desirable to keep the family farming unit the main production unit on the farm and through the harsh times to be able to keep them there so that they don't disappear and we end up with the farming systems like a lot of the European countries have, where there is absentee landlords and very large landholders, and the people, the majority of the people who work the land are merely tenants. (Male, Executive Member [1], CUA)

> We don't want to be involved in or dependent on hand-outs from government. We would much prefer to be self-reliant. But to ask people what they have gone through in what they have, in the last seven or eight years, in terms of the drought and low commodity prices, to all of a sudden be self-reliant, is just beyond imagination. So we do support self-reliance, but we argue very strongly that you can't expect people who are on their knees to all of a sudden be self-reliant without some sort of transition process, and assistance measures to get them through that. (Male, Executive Member, UGA)

The previous statements show the different rationalities through which Queensland-based producer organizations constitute self-reliance. While these organizations support self-reliance, this is constituted as part of a *lifestyle* in which the management of risk requires social forms of assistance. Here, producers are constructed as part of a wider collectivity of family farmers whose capacities to manage effectively can be addressed through the mutualization of risk. Thus, farm organizations constitute self-reliance in a way that emphasizes the problems and contingencies of rendering farm viability calculable through formal business techniques. This issue is developed further in Chapter Five where the rationalities underpinning farmers managerial skills are examined.

CONCLUSION

This chapter has argued that self-reliance represents an assemblage of rationalities and technologies for governing the conduct of farmers in an advanced liberal way. The individualization of risk is perhaps the most important aspect of self-reliance. Both the physical and economic environment are constituted as individual risks that can, through the application of business management techniques, be managed effectively for increased financial returns. Technologies of business management have attempted to reconstitute farm viability in a way that makes it amenable to financial calculation, and thus render the practices of farmers visible to authorities. This, in turn, enables the formal economic capacities of farmers to be constructed as the legitimate sites for reform and adjustment. This is most obvious in the promotion of property management planning that seeks to specify the practices and techniques for operating a viable property. In this way self-reliant farmers are effectively 'produced' through their use of particular business management techniques. Those producers who do not adopt a responsible business-oriented relation to risk are constituted as *targeted populations* who are seen to lack the capacities for managing their farms in a viable manner. Such individuals are deemed to be candidates for welfare and provided with forms of knowledge, such as financial counselling, to indicate that their properties are unviable in the long-term.

While a self-reliant approach to the governing of agricultural change is represented by authorities as a necessary means of improving the economic efficiency and sustainability of the farm sector, this strategy has been possible only through a heterogeneous ensemble of knowledge and events. Thus, the previous association of self-reliance with farming as a way of life, the constitution of sustainability as an individual management problem, the questioning of drought as a disaster, the continuing period of dry weather in eastern Australia and the continued discourse of international competitiveness have all been important in the emergence of self-reliance as a means of restructuring the conduct of farmers.

As self-reliance is a heterogeneous assemblage of elements for shaping the capacities of farmers, it follows that there exists any number of sites at which the objectives of authorities might face practices that attempt to re-constitute these strategies, or are altogether oppositional. While issues of exceptional circumstances, the definition of viability, and the lifestyle aspects of farming were raised briefly in this chapter, the following chapter examines in more detail some specific sites that are significant in defining the governable limits of managed risk.

Chapter 5

SELF-RELIANCE AND THE LIMITS OF MANAGED RISK

The promotion of self-reliance at both a government and non-government level involves advanced liberal rationalities and technologies of governing that attempt to constitute primary producers as economically rational risk managers. While such a strategy is constituted discursively as a logical, coherent and truthful means of governing agricultural change, Chapter Four showed that this is subject to a degree of contestation, particularly by State administrative authorities and some farm organizations. This chapter expands in more detail on how advanced liberal assemblages of governing agricultural change are shaped. Three sites are examined at which rationalities and technologies of managed risk are problematized in governing the conduct of farmers – drought exceptional circumstances, the local knowledges of farmers and environmental sustainability. In exploring these sites, the central point of interest is how advanced liberal techniques of rule are called into question, and the consequences of such problematizations for re-politicizing the regulation of agricultural change. Prior to an analysis of the three sites, it is first necessary to provide a background outlining how the consistent 'failure' of the Rural Adjustment Scheme to achieve its stated objectives is a crucial consideration in understanding the limits of attempts to 'restructure' farmers' conduct.

MANAGING AGRICULTURAL CHANGE: A CONGENITALLY FAILING OPERATION

Since their establishment in the early-1970s, programs seeking to manage agricultural change – such as the Rural Reconstruction Scheme and Rural Adjustment Scheme – appear to have rarely achieved their desired objectives. According to statistics, large-scale autonomous changes have continued to take place outside the boundaries of these programs. In fact, it has been estimated that between 1985/86 and 1994/95 there was a 1.3 per cent decline per year in the total number of Australian farms, with the largest changes occurring in broad-acre farming and dairying (ABARE, 1997). This translates into a drop from 130,000 to 115,368 farms.

Statistical representations of the farm sector, such as those noted above, have been used as the basis for arguments by economists that the rate of exit by farmers from the industry is not occurring in the most efficient or desirable manner. Musgrave (1990), for instance, in support of this view, estimated that in any one year less than two per cent of farmers applied successfully for assistance through the Rural Adjustment Scheme with only a small proportion of those actually leaving the industry. Between 1977 and 1987 this meant that less than one quarter of successful RAS applicants left the industry as a result of receiving adjustment assistance (Musgrave, 1990:257). Figures from the *Mid-Term Review* of the 1992 Rural Adjustment Scheme suggest that up to 18 per cent of producers applied successfully for RAS assistance between 1992/93 and 1995/96, with only one per cent of these successful applications being for re-establishment grants (McColl et al., 1997). However, it is important to note that this was a period of severely dry weather throughout eastern Australia with exceptional circumstances support comprising just over half the total assistance (see Commonwealth of Australia, 1994). In summary, between 1977/78 and 1997/98 there were only 3,081 approvals for RAS re-establishment grants (Commonwealth of Australia, 1994; and Rural Adjustment Scheme Advisory Council, 1995-1998). The above figures have enabled the construction, by economists, of a farming sector that is argued to require better governance in the interests of improved resource efficiency.

Interviews with Commonwealth public servants suggest also that the RAS was ineffective in producing the desired changes in farm structure. As two of these public servants argued:

> There is very little evidence to suggest that RAS has had much impact at all on rural adjustment. The McColl Review [Mid-Term Review of the 1992 Rural Adjustment Scheme] suggested that the productivity improvement that had taken place, it would be very difficult to say it wouldn't have taken place anyway. On re-establishment...it is questionable whether the incentive in the re-establishment provision is so great that it makes a meaningful contribution. (Male, Assistant Secretary level, DPIE)

> I think any positive effect...[from the RAS]...would have been pretty limited, because I think it's been there with much bigger events pushing part of it. It's not stopped the rate of movement out of agriculture, it's not helped a great deal of farmers, and one of the things the McColl Review said was that there were ways of managing exits in a more positive way – in a way that causes less distress to farmers (Male, Middle Management level, RIRDC).

These statements draw attention to governing as a 'congenitally failing operation' (Miller and Rose, 1990:10), a key feature of liberal forms of governing. Here, the mentalities of government are 'eternally optimistic', but the practice of governing involves heterogeneous and contradictory elements with 'the solutions for one program tending to be the problems of another' (Miller and Rose, 1990:10-11). Obviously this does not mean that programs such as the RAS are a failure in some absolute sense. If this were the case, it is unlikely that the RAS would have persisted for such a long period of time. What is important is how the viability of some farmers has been constituted by expertise as an ongoing problem, seen to require the intervention of authorities.

The seemingly low rates of exit noted above have long been used by academic economists and public servants as a justification for continued government involvement to increase the number of 'unviable' producers leaving the sector, and to redistribute agricultural resources toward their most efficient use (eg. Kingma and Samuel, 1977; Campbell, 1980; Musgrave, 1990; Gow, 1996; Hyde, 1997). However, attempts to facilitate change in the conduct of farmers is represented as an ongoing issue of concern, evidence of which can be found in the number of times the RAS has been modified since its introduction. Such problems are frequently traced to the nature of the programs themselves. A number of RAS reviews (eg. Industries Assistance Commission, 1984; Coopers and Lybrand WD Scott, 1988; Synapse Consulting, 1992) argue that rural adjustment programs are inefficient for not encouraging in the most optimum manner acceleration of unviable producers out of the industry, while maintaining those deemed as viable. More specifically, criticisms since the 1990s have focused increasingly on how the RAS encourages dependence on government assistance and inhibits a self-reliant approach to farm management (Synapse Consulting, 1992; Commonwealth of Australia, 1994; McColl et al., 1997).

Extending this argument, Stayner (1996) suggests that adjustment programs have in some circumstances been used as a 'convenient vehicle for directing industry assistance to the farm sector at times of acute stress' (187). As one public servant noted in support of this view:

> You could mount an argument that RAS has inhibited adjustment. I think that there could be a fairly strong argument mounting for that in time because I do think that, in fact, what has happened with RAS is that there have been times when RAS has been used very successfully as a political 'out' for hard decisions. (Male, Branch Director level [1], DPIE)

The previous statement emphasizes that governing is a contested exercize in which programs frequently fail to meet their original objectives and, in fact, can be shaped to accord with a different set of purposes. This accords with the arguments of Foucault (1982) who notes that every power relationship implies a strategy of struggle: 'Each constitutes for the other a kind of permanent limit, a point of possible reversal' (225). Genealogically, these struggles are constitutive of historical attempts to make practices of governing thinkable and manageable. Such struggle is crucial in drawing attention to how particular forms of rule are assembled, and the forms of reason that seek to (re)constitute and shape rationalities of managed risk.

Rather than engage in an already well-rehearsed and protracted debate concerning the extent to which continued intervention through the Rural Adjustment Scheme either increases agricultural efficiency, or keeps those producers defined as unviable on the land, it is more fruitful to the purposes of this chapter to examine the sites through which 'managed risk' – as an assemblage of rationalities and technologies for governing in an advanced liberal way – encounters limits. What is of interest here sociologically is not whether farm adjustment programs have, or have not, been successful in any absolute sense, but the specific problems raised from governing in this way. Focusing specifically on the period 1990 to 1997, three examples are used to show how limits are imposed on rationalities of managed risk at both a formal political level, and at a local farm level: exceptional circumsatnces provisions, the 'local' knowledge of drought affected producers in Queensland, and the discourse of sustainability. An examination of these three sites is useful in showing how governing never fails as such but, through competing and sometimes contradictory political rationalities, encounters limits and a (re)politicization of advanced liberal rule.

'Exceptional Circumstances' as a Site of Contestation

As noted in Chapter Four, the *National Drought Policy* changed the way in which drought assistance was delivered. There was a renewed focus on drought as a 'manageable risk' rather than as a 'natural disaster' with assistance distributed via the Rural Adjustment Scheme as opposed to the Natural Disaster Relief Arrangements. However, a rationality of governing drought as a disaster was not completely abandoned in 1990, but continues – through Exceptional Circumstances provisions – until the time of writing (late 2002). The aim of exceptional circumstances is to 'provide short-term targeted assistance for rare and severe events outside normal risk management strategies' (Rural Adjustment Scheme Advisory Council, 1996:25). According to the legislation, interest rate subsidies of up to 100 per cent are available for productivity improvements, carry-on finance and debt restructuring in circumstances defined by the Minister as exceptional.

These provisions have proved to be remarkably durable over time remaining in a form virtually unchanged since 1992. When the Rural Adjustment Scheme was abolished by the conservative Liberal/National Coalition Government in 1997, Exceptional Circumstances became a program in itself as part of the *Agriculture - Advancing Australia* ('Triple A') rural policy package. Its provisions were expanded to include a Drought Relief Payment, introduced originally in 1994, and possible eligibility for additional social security support (Dept. of Primary Industries and Energy, 1998d). Interest subsidies of up to 100 percent were retained. Since its introduction, exceptional circumstances provisions have remained the most widely used form of support out of all the programs in the former Rural Adjustment Scheme. This is particularly the case in Queensland and New South Wales where, during the years 1992 to 1996, over 50 percent of total approvals under the RAS were for exceptional circumstances support (McColl et al., 1997:56). While there may have been a genuine need (however that may be defined) for such assistance, one would assume that if state agencies were truly committed to a self-reliant approach to managing risk, applicants in difficult circumstances would simply be told that they are unviable and their applications rejected. How then might one account for the continuation of exceptional circumstances?

Exceptional circumstances represents a fairly explicit means of assistance that seems to be generally inconsistent with the objective of managed risk promoted through rural adjustment policy. It could easily be argued that exceptional circumstances provisions represent little more than a political 'safety-valve' that serves to protect the legitimacy of governments when faced with circumstances that cannot be resolved through market-based measures. However, this assumes that exceptional circumstances support can be drawn upon at will without having an effect on advanced liberal rationalities of rule. The widespread use of exceptional circumstances in both Queensland and New South Wales indicates that this is not the case. These provisions have indeed shaped how a rationality of managed risk is constituted, and particularly the types of farmers seen as having the capacities to maintain viability. On this basis, exceptional circumstances provisions are not the polar opposite of manageable risk, but represent a strategy for ordering and shaping the limits of advanced liberal forms of rule. In other words, it as an important means of contesting strategies of managed risk.

Many scholars argue that governing is seldom a straightforward matter of implementing a unified program of rule. There are always contradictions, contestations and negotiations.

However, rather than conceptualizing these as external obstacles to rule, they need to be seen as a constitutive part of the governing process (O'Malley et al., 1997). Drawing on the work of Michel Callon and Bruno Latour, Rose (1999) labels this a process of *translation*. While I do not examine here the complex processes through which exceptional circumstances provisions were translated by authorities, I seek to analyse its effects manifested in the continued debate over the extent to which drought is manageable. The analysis of government documents, as well as the use of semi-structured interviews conducted with Commonwealth and State public servants and representatives of farm organizations, is used to show how exceptional circumstances provisions impose limits on those strategies that seek to govern through the individual managerial capacities of farmers.

In the last ten years there has been considerable debate among both public servants and farm organisations concerning exactly which droughts should be classed as exceptional. While not defining clearly what exactly should be considered an exceptional drought, the view of a number of senior Commonwealth public servants was that most climatic variability should be regarded as normal. For example:

> The difficulty with interventions to assist a sector through difficult times is they also create a moral hazard in the insurance sense, and you can see that in the sense that, basically the government underwrites farmers in exceptional circumstances (Male, Assistant Secretary level, DPIE).

> I have a bit of a difficulty with the concept of [exceptional circumstances] in terms of other businesses are intended to be self-reliant, and why do farmers get preferences? If all businesses went broke there would be a social problem also. (Female, Senior Officer level, DPIE)

> I think that the expectation that farmers should be able to put into their risk management strategies the concept of one in a hundred [droughts] is reasonable, I haven't got any problem with that. I think that there has to be realism, and unless there's a major catastrophe I don't see why the government should get involved. (Male, Branch Director level [1], DPIE)

Such views are characteristic of most policy documents on drought in that they emphasize that farmers should take an active and personalized approach to managing climatic variability, rather than relying on government support. They attempt to juxtapose self-reliance to government assistance, with the former seen as most desirable and efficient.

Nevertheless, as argued in Chapter Four, self-reliance does not mean a lack of regulation by authorities: 'self-reliant' farmers are effectively produced through advanced liberal rationalities and technologies of governing. Here, self-reliance is not some natural ideal to be aspired to, or an absence of state intervention, but a strategy within which the capacities of farmers to manage risk are constituted. Self-reliance, as a lack of government intervention is thus something of a misnomer. This was well illustrated by one respondent who noted that self-reliance would be possible only if markets operated in an ideal-typical manner:

> A well operating free market depends on the fact that there are no externalities, there are no spillover effects, there are no failures in terms of market prices to signal the appropriate decision or purchase to be made… If your market is subject to those sort of

> distortions, you know, and if it's a large proportion, then self-reliance doesn't mean a thing. (Male, Branch Director level [3], DPIE)

Given that the delivery of farm adjustment and drought assistance is based on the argument that markets left to themselves are incapable of delivering the most efficient outcomes, this respondent suggests that self-reliance requires some form of state intervention. Most interviewees acknowledge that a role for governments is always necessary in enhancing farmers' capacities to manage their enterprizes and market their commodities. This accords with advanced liberal technologies of governing where authorities attempt to enhance economic efficiency through 'empowering' individuals in the art of active self-management (Rose, 1993; 1999).

While endorsing the importance of self-reliance as an objective involving intervention by state agencies, many public servants challenged the construction of farming risk as entirely manageable, arguing that this ignored a range of external conditions. For example, one participant raised the issue of political support for governments to respond to severe events seen to be outside the control of farmers:

> ...if you look at the last big drought in '93 and '94 there were millions of dollars raised in cities by ordinary people who wanted to help. If that happens, with people putting their hands in their pockets to help, then governments are not going to ignore that sort of sentiment. Therefore, Australian people are saying, 'yes, we want to have some sort of response to these sort of severe events'. All the theory in the world is not going to win against that sort of response. (Male, Senior Officer level, DPIE)

Other participants suggested that the existence of risks beyond the everyday control of farmers compromized seriously the capacity of these producers to manage. For example:

> The rural industry is different to most other industries. There is no control over the elements. [Farmers] have probably little control over the prices for their products, they cannot set the price that they are going to sell their crop at... I think the rural industry is somewhat unique and that is why governments have a role to play and have done so. (Male, Middle Management, QRAA)

> I am not sure whether...[government]...has a greater obligation to farm families than it does to urban families, and I am still trying to resolve that in my own mind about whether you can really manage or whether you should be able to manage effectively for climate, and whether we have just a lot of people who can't, or whether there are things that are just beyond which you can't just expect people to manage. (Female, Senior Management, QDPI)

> ...exceptional circumstances is always going to be a difficult area no matter how self-reliant farmers are. There's always going to be a set of circumstances which comes in and is beyond the ability of any farm business, or any business for that matter, to provide for. (Male, Executive level, NFF)

The statements outlined above show that while self-reliance remains a stated policy objective in governing drought at both the Commonwealth and State level, and among farm organizations, its potential to manage climatic variability remains subject to contestation. This questions the argument advanced by many rural sociologists that neo-liberalism represents a hegemonic 'project' that leads to a widespread loss of legitimacy for socialized forms of regulation (see Lawrence, 1987; McMichael and Myhre, 1991; Bonanno, 1994; McMichael, 1996; Campbell and Lawrence, 2000; Gray and Lawrence, 2001). In fact, skepticism concerning advanced liberal forms of governing is here significant in creating a site of contestation through which managed risk can potentially be shaped in a distinctively social manner.

The political recognition that there are some circumstances for which farmers can not be expected to reasonably manage has led to attempts to more objectively quantify exceptional droughts, in order to delineate clearly the types of conduct that comprise 'normal' management strategies. This has involved a process of defining, linking and measuring in which the knowledge of expertise such as scientists, meteorologists and economists is called upon to render heterogenous elements into a technical and calculable form. Such data is then drawn upon by state agencies to assess whether farmers are in need of exceptional circumstances support. Since 1994, six core criteria have been used by the Agricultural and Resource Management Council of Australia and New Zealand (ARMCANZ) – a collective of Commonwealth and State agriculture and resource management ministers – to assess whether a drought is exceptional. These are:

- Meteorological conditions (the primary condition for the existence of severe drought).
- Agronomic and livestock conditions.
- Water supplies.
- Environmental impacts.
- Farm income levels.
- Scale of the event. (derived from Lembit and Kingma, 1995)

Despite the existence of these criteria, the definition of an 'exceptional' drought remains open to debate with the views of state agencies not always according with what farmers and graziers believe to be a severe drought (see Stehlik et al., 1999). In fact, there have been a number of cases where differing definitions of 'exceptional' among producers, State politicians and Commonwealth ministers, has led to heated debates in both Queensland and New South Wales (eg. *Rural Weekly*, 1999). Nevertheless, economists have continued to push for more objective indicators to define exceptional drought so that government decision-making processes can be standardized and the characteristics of 'normal' farm management more clearly delineated (see White and Bordas, 1996). Self-reliance, as a strategy for transforming farmers from recipients of assistance to active agents in the management of farm problems, has not then been abandoned. However, neither has the seemingly contradictory idea that farmers are entitled to some form of assistance in times of severe stress. This provides evidence for what Stenson and Watt (1999) call *hybrid* practices of governing, where both social and advanced liberal rationalities of rule are assembled within programs. It seems likely that the issue of severe drought will continue to remain important in shaping rationalities of managed risk and in defining the 'proper' managerial capacities of farmers.

'Local Knowledge' as a Site of Contestation

Drought forms an important site of contestation over rationalities of managed risk, not only at the formal political level, but also at the level of farmers' understandings of farm management. While it is conceptually problematic to draw strict distinctions between 'expert' and 'local' knowledge (see Murdoch and Clark, 1994), farmers' understandings of their practices are frequently viewed by expertise as obstacles to the rational application of scientific knowledge to agricultural environments (eg. Kloppenburg, 1991; Long and Long, 1992; Vanclay, 1992; Portela, 1994; Clark and Murdoch, 1997; Millar and Curtis, 1999; Higgins, Lockie and Lawrence, 2001). In fact, the principal reason why State departments of agriculture have historically employed extension officers is precisely to influence the practices of farmers in ways that are consistent with the objectives of these agencies (for an Australian example see Vanclay and Lawrence, 1995). It is perhaps therefore not surprising that explanations for the apparent failure of government programs to influence farm practice are aimed at farmers themselves, as much as the practices and calculations of state agencies. Such explanations tend to focus on first, a lack of awareness or misunderstanding by producers (see Commonwealth of Australia, 1995; Garnaut, Robinson and Lubalwa, 1997); second, the belief by producers that they do not need these programs (Commonwealth of Australia, 1995) and third, a lack of information by producers concerning the financial position of their farm, alternative ways of using the farm to generate income, or alternative job opportunities (McColl et al., 1997). Within all these criticisms, it is the knowledge of farmers that is seen to provide the obstacle to more efficient governing of farm practices.

One of the most comprehensive works in Australia to date exploring the influence of local knowledge on the individual management of agricultural change is Stayner's (1994) economic research on farm adjustment in the Bland Shire of New South Wales. This project found that adjustment actions taken by farmers, assisted through the Rural Adjustment Scheme, were much more varied than simply 'getting big or getting out'. Stayner (1994) found that family-related factors were essential in determining farm management, decision-making and desired outcomes from the enterprize, and did not correspond neatly with the assumption made by many economists of adjustment actions being driven primarily by financial goals (see also Gow and Stayner, 1995; Stayner, 1996). In fact, he states in a more recent article that many farm families appear willing 'to absorb quite large reductions in household income, at least in the short term, to preserve the continuity of the farm business and way of life' (Stayner, 1996:173). Subsequent Australian studies have found, like Stayner, that a complex range of financial and non-financial goals shape farm management decisions (Bamberry et al., 1997; Kilpatrick, Johns, Murray-Prior and Hart, 1999). From the perspective of economic expertise, the fact that non-financial goals play a part in farm decision-making would be considered highly irrational requiring stronger incentives so that producers know when to exit the sector.

Stayner's (1994) research is clearly important in recognizing that producers view their properties as more than just business enterprizes – a finding that represents a challenge to programs encouraging managed risk. However, while cultural and lifestyle considerations are taken into account – which could be argued to be a significant step in itself – this is for the ultimate purpose of the New South Wales Rural Assistance Authority to govern farm adjustment in a more efficient manner. Moreover, Stayner assumes, like many economists, that the need for adjustment is a self-evident economic fact requiring farmers to act in ways that support rather than impede inevitable economic processes. Rather than representing the practices of producers as disruptive to change, the approach taken here argues that local farming knowledges play an important role in rendering farm practices intelligible as a governable site. This draws on O'Malley's (1996) argument that local knowledges are crucial in making the operation of governing at specific sites possible. Analysis of selected data from a recent study of drought (Stehlik et al., 1999) shows how producers constitute planning practices, and how such understandings differ from the objectives of those programs seeking to govern through strategies of managed risk.

The study, funded by the Rural Industries Research and Development Corporation (RIRDC), and entitled *Drought in the 1990s: Australian Farm Families' Experiences*, involved comparing the experiences of farm families in two regions affected by drought (1989-1996) – one in Central Queensland (beef producers), and the other in the Riverina region of New South Wales (sheep/wheat producers). Fifty-five male and female adults were interviewed in Central Queensland and 54 in New South Wales using semi-structured interviews. Given the constraints of space, the series of interviews with the 55 Central Queensland producers is of most interest in this chapter. Seven questions were selected from the interview schedule that related to the ways in which farmers manage their properties. These questions are grouped below according to five assumptions that are rarely questioned in policy documents seeking to better govern agricultural change. The purpose of selecting and categorising questions in such a way is to demonstrate that farmer knowledges are useful in understanding why programs might not always achieve their desired ends.

Assumption One – Individual managerial ability is the most important factor in effective drought management

It was necessary, as a first step, to establish the importance that producers attached to property management. Of the 52 producers who responded to the proposition: *How strongly do you agree with this statement? Management really makes a difference in drought*, 61.5 per cent strongly agreed, 34.6 per cent agreed, and only 1.9 per cent disagreed. Table 5.1 provides a summary of these responses.

Table 5.1: Management Really Makes a Difference in Drought.

		Valid %
Strongly agree	32	61.5
Agree	18	34.6
Disagree	1	1.9
Strongly disagree	0	0
No opinion	1	1.9

N=52

The importance of management was confirmed by responses to the question, *Are there some farmers who are recognized as better managers of drought in this area?* of which, out of 54 responses, 87.0 per cent answered ‘yes’ but only 13.0 per cent answered ‘no’. Table 5.2 provides a summary of these responses.

Table 5.2: Are there some farmers/graziers who are recognized as better managers of drought in this area?

		Valid %
Yes	47	87.0
No	7	13.0

N=54

These figures suggest that an overwhelming majority of respondents believed some form of management capacity and skill to be crucial. Nevertheless, this did not necessarily mean that formal managerial skills were viewed as a prerequisite to effective management. The question, *How are they (the best managers of drought) recognized?* elicited a number of responses, only one of which mentioned formal education. Conservative stocking and financial resources were two of the most common responses. For example:

> They are only better managers because they probably have a lot less stocking rates and because they don’t have debt commitments (Male Producer).
> I think they are assisted by the fact that they are not financially strapped. They are not worried about debt, or their debt structure is such that they can go further into debt to make some improvement to their fencing (Female Producer).

However, some producers emphasized the fine line between good management and luck. For instance:

> People who have sold...about 50 per cent of their cattle in the beginning have turned out to be better managers. Whether that was good luck or good management I don’t know (Male Producer).

> I’ve known some people, their conventional strategy is [to] shut down until it is over. If they have got money behind them they can do it. Others battle on [to] feed their stock, [and to] get through. [If] It rains, they breed up and they are sitting pretty. It varies a hell of a lot and it depends on how the cookie crumbles and so on (Male Producer).

These statements indicate that the possession of formal managerial skills is seen as but one element in effective property management, with financial resources generally considered as most important. However, in some cases an element of luck was seen to shape the types of responses that producers made to their environment.

This accords with an Australian study by Bamberry et al. (1997) who argue that ‘good’ farm management is defined by farmers according to the managing of family relationships in terms of the business structure, personal attributes and a sound capital base. Moreover, informal management skills and ‘learning on the job’ are frequently seen as more significant than formal managerial qualifications. As Bamberry et al (1997:143) state, ‘The breadth of scope of farm management makes it difficult to construct precise measures of good farm management’. Such findings represent a direct challenge to the objectives of farm adjustment and drought programs in which formal business management is argued to be the most 'truthful' means of measuring viability.

Assumption Two - Business plans are an essential tool of 'good' farm management

Once the role of management in both male and female producers' on-farm strategies was established, their possession of a business plan was ascertained. This issue was addressed through the question, *Do you have a business plan for the property?* From a total of 54 responses, 44.4 per cent of producers stated that they had a business plan, while 55.6 per cent did not. The responses are detailed in Table 5.3.

Table 5.3: Do you have a business plan for your property?

		Valid %
Yes	24	44.4
No	30	55.6

N=54

These results should, however, be treated with caution. Lack of a business plan did not necessarily mean having no plan at all for one's property. Many of those respondents who answered 'no' actually had plans, but just not those recognized by the experts such as accountants and farm business advisers. As one respondent noted:

> Everybody has got a business plan, some have got it in their head, and some have got it on paper (Male Producer).

Similarly, another remarked:

> It is basically all in my head, I don't know what they call a business plan... There is very little you can do in the way of a business plan here except to develop and get your debt down and get something off farm, an investment off farm. (Male Producer)

A female producer noted the importance of income in implementing a 'proper' business plan:

> [We have]...a very vague one. Not formal. We don't and we should...We talked about it a bit. At the moment we feel we need a few dollars to make the business plan work. That is wrong but that is the way you go. (Female Producer)

The above statements indicate that while many producers agreed that having a business plan was important, their interpretation placed less significance on formal written technologies of planning. Formal business plans are a technique emphasized in programs such as the Rural Adjustment Scheme, the Property Management Planning Campaign and FarmBis, and are increasingly a pre-requisite sought by financial counsellors prior to the granting of financial assistance. In fact, formal planning is represented in official policy documents as the most truthful means through which farmers can prove their viability. Those producers who do not possess a formal plan are thereby constituted as unviable or counselled to formulate one in the interests of prudent financial management. Lack of a formal plan may be interpreted by the close relationship between farmer's business and family goals (see Gow and Stayner, 1995) where the latter are more difficult to quantify formally. Thus, while producers in the drought study saw planning as important, their desire to have a basic level of income in order to remain on the land provided a potential site of contestation to more formal technologies of calculation. The fact that this may have excluded them from receiving assistance, or categorized them as unviable altogether, demonstrates the significance of non-formal techniques of farm management.

Assumption Three - Farming should be treated solely as a business enterprize

The relationship of producers to farming was further explored through the question, *To you, what are the important things about life on the land?* It was argued in Chapter Four that contemporary rationalities for governing farm viability increasingly emphasize that farmers should make decisions based primarily on profit-making considerations. However, responses from the farmer interviews pointed towards a range of other non-economic factors in influencing how producers understood their decision-making environment. *All* respondents noted lifestyle reasons as the most significant factors for remaining in the industry. Common responses for remaining in farming included the benefits of independence, being one's own boss, freedom, and being close to nature. More specifically, respondents noted:

> When I worked in town I used to love my work, but it was very boring. So it is a challenge. I definitely like the challenge. You are your own boss. The fact that what you do you are accountable. You can't really blame anyone else. (Female Producer)
>
> Freedom, independence. I wouldn't be able to live in a city, I'm sure I wouldn't. (Male Producer)
>
> You certainly learn to be independent. I grew up in the family business, I am part of it. (Female Producer)
>
> I guess that from my point of view, having worked on wages, being your own boss is important and having that personal challenge. Knowing that if you have done something well or succeeded in something it is because you have done it. There is also a lot of personal trauma. It is without a doubt the best place to bring up a family. They can do a lot of things here... I guess I enjoy the lifestyle. (Male Producer)

These statements support previous Australian sociological research where independence, hard work, being one's own boss, and worthiness of farming as an occupation were found to be important in owning and operating an agricultural property (Gray, 1991; Kidman, 1991; Poiner, 1991). Also of significance is the opposition between city as bad and rural as bringing out the best in people, which corresponds with research carried out by Dempsey (1990), Kapferer (1990), Poiner (1991) and in Aitkin's (1988) reflections on countrymindedness. However, it is also important to note that these values come at a cost in terms of high stress levels and poor health (see Gray, Dunn and Lawrence, 1993; Cheers, 1998).

The perspectives presented above support Stayner (1994) and Gow and Stayner's (1995) argument that lifestyle and personal goals are crucial influences in affecting the decisions that producers make concerning the management of farm viability. This is supported further by Chamala, Coughenour and Keith (1983) who found that in a survey of farmers on Queensland's Darling Downs, 71 per cent agreed or strongly agreed that profit and capital gain were only a small part of their overall commitment to farming. More recent research (Bamberry et al., 1997) suggests that profit and capital gain are important, not as ends in themselves, but as an integral part of long-term farm development in which a variety of other non-economic factors also play a significant role. It appears that farmers' attachment to farming as a lifestyle remains important in structuring their everyday practices. In this respect, such practices may be viewed by economic experts as disruptive to the adoption of more rational risk management strategies. Nevertheless, failure to incorporate these local knowledges may contribute to many farmers viewing advanced liberal rationalities and forms of calculation as contradictory or otherwise irrelevant to their everyday experiences (see O'Malley, 1996).

Assumption Four- Drought can be managed like any other risk

Once the relationship of producers to their management practices was ascertained, it was then important to examine the effect that this had on their constructions of risk. Risk is conceptualized here not as a 'reality' but as a way of ordering reality and rendering it into a calculable form (Dean, 1999:177). That is to say, for a phenomenon to be considered a risk, an amalgam of calculations and knowledges must be deployed. For instance, if producers constructed their properties primarily as a business, then one might expect that they would see their risk as manageable through business techniques.

Official government documents claim that drought is a phenomenon that should be managed through formal techniques of farm planning and management (see Commonwealth of Australia, 1990; 1992b; 1995). The question, *What is your opinion of the view that government should not consider drought to be a natural disaster or an unusual event and therefore not provide drought assistance to farmers?* sought to elicit producer's constructions of environmental risk, and whether they saw drought as something which could be managed according to normal risk management strategies. The overwhelming response from producers was that drought was more a natural disaster that needed to be responded to through *public* forms of support rather than something that could be managed individually. For example:

> It is a natural disaster. It sneaks up on you, it doesn't grab you real quick. (Male Producer)

> It should be a natural disaster. You can't get any more natural. It is like someone up there turned off the tap. (Female Producer)

A number of respondents noted limitations in managing for drought, particularly the influence of cashflow and market conditions:

> Well it is certainly partially manageable. I think...the government should encourage people to become drought resistant... But you can't manage for drought for six years. (Male Producer)
>
> Drought is not completely manageable alone. I think if people are going to try and be drought-proof it is going to cost a lot of money, and when there is not rain and not grass in the paddock, well, there is nothing you can do. (Female Producer)
>
> When you talk about drought, drought is drought. Drought itself is not the problem, it is cash flow... If you get the two come together you have got a big problem, as we have had in the last 12 months. We've had a dry spell, a drop in commodity prices, those two together is disastrous. If you have got a drought, and a good return for your product it is a lot less of a problem. (Male Producer)
>
> I think it is a natural event to a degree. But this is a low profit margin that we receive now, it is extremely hard for farmers to be able to put money aside to help in those dry times. The market makes it worse. I think we will always have droughts. It is just hard to be able to put...[money]...aside to be able to cope with. (Female Producer)

These statements draw attention to the apparent limitations of using formal management plans when risk is constituted as a phenomenon seen to exist beyond the managerial capacities of producers. The intervening factors that are seen to make drought unmanageable include length of period under drought conditions, low commodity prices and lack of financial reserves. While programs such as the recent Farm Management Deposits Scheme (FMDs), and the older Income Equalization Deposits Scheme (IEDs), have sought to encourage the building of financial reserves so that producers can cope during times of drought, this appears to be of little use if prices are low or if conditions are such that income simply cannot be put aside. Exceptional circumstances support is available for producers who do not have sufficient reserves. However, the political rationality underlying such support is that, with the optimum economic environment, producers will be provided with the opportunities to build up cash reserves and be self-reliant in the long-term. For some producers, it is questionable whether these claimed opportunities will redefine environmental risk in a way that is consistent with their constructions of farming practice as a highly variable and, at times, unpredictable exercize.

Assumption Five - Commodity prices should not constrain risk management strategies

While producers identified a number of limitations in the management of environmental risk, it was also considered important to gauge their attitudes towards financial risk, and whether this was something that was seen to be manageable through individual managerial skills alone. The question, *Has your ability to manage during the drought been affected by commodity prices?* sought to examine this issue. Out of 46 responses, an almost unanimous 97.8 per cent agreed that commodity prices had constrained their management strategies and practices. This is represented below in Table 5.4.

Table 5.4: Has your ability to manage during the drought been affected by commodity prices?

		Valid %
Yes	45	97.8
No	1	2.2

N=46

These figures suggest that commodity prices have a highly significant impact on the ways in which producers understand and therefore plan for and manage climatic risk. Thus, while producers may want to take a more personalized approach to risk, rapid changes in commodity prices might wipe out the financial resources that are deemed as necessary to manage in a more formalized manner. This supports the findings from the previous question where many producers questioned a more planned approach to management as inconsistent with their constructions of natural environments as variable and requiring flexible forms of planning.

Drought Management: An Evaluation

As noted in Chapter Four, the rationalities and technologies promoting self-reliance seek to constitute farm practice as something that, if managed using particular business techniques, enhances producers' profitability in a sustainable manner. However, this is clearly not in accord with what many producers in the drought study 'knew', and was not seen as practical to their immediate circumstances. While not directly contesting advanced liberal rationalities and technologies of governing, producers politicized their practices as more complex than those defined through the expert discourses of manageable risk. Indeed, management from the farmers' perspectives emphasized practices that were intuitive, experiential, variable and unquantifiable. These stand in quite sharp contrast to those practices promoted through discourses of managed risk.

While the evidence presented here does not seek to draw a link between programs of managed risk, and their effects on farmers' management practices, the material certainly points toward a high probability of producers acting in ways that do not accord with advanced liberal objectives. This does not mean, however, a failure on the part of farmers to manage their properties, as implied by some government documents (eg. Commonwealth of Australia, 1995). The idea that drought needed to be managed was not in dispute from the perspective of the interviewees. What was politicized by producers was the construction of drought as something that could be managed in a quantifiable and formal manner. Most producers recognized that some degree of management ability was needed in order to effectively run a viable property during drought conditions, but that such ability depended on a variety of factors. What this example shows is that local knowledges do not always neatly accord with governmental objectives, and that consideration of these knowledges is important in showing how particular programs seem to constantly fail in achieving their original desired goals.

'SUSTAINABILITY' AS A SITE OF CONTESTATION

The issue of environmental sustainability represents a third site through which contestation over rationalities of managed risk can be observed. Government documents claim that through the encouragement of farmer self-reliance, those producers defined as viable will become more environmentally responsible by amalgamating land and engaging in formal property planning practices, while those who are defined as unviable will know that their management techniques are unsustainable, and leave the sector. However, as noted earlier in this chapter, this does not appear to be occurring in practice. Many farms judged by authorities to be poorly performing are, in fact, choosing to earn off-farm income in order to remain on the land – and therefore stay 'viable' – until the point of inter-generational transfer (Barr and Ridges, 1998). While it could be argued that such stubbornness and seeming irrationality on the part of certain producers is contributing to family poverty, and increased land degradation, through practices as overstocking, the current focus on individual management as the source of problems – and therefore the site to administer solutions – de-limits considerably the types of legitimate responses that can be pursued. As noted in Chapter Four, this has much to do with definitions of viability that equate sustainable land management with formal business planning, economic efficiency and productivity.

The perceived failure of many producers to respond in the manner desired by governments has served to draw attention to the ways in which state agencies are dealing with adjustment issues and managing agricultural change (see Synapse Consulting, 1992; Commonwealth of Australia, 1994; Commonwealth of Australia, 1995; McColl et al., 1997). While self-reliance continues to be endorsed by both public servants and farm organizations as the best means of achieving farm productivity, its capacity to address issues of sustainability remains subject to contestation. This, to some extent, politicizes individual managerial skills as a legitimate means of governing farm practices. Thus, if environmental issues are seen as public political problems, rather than non-political individual management problems, this increases the scope for debate on strategies of expenditure and monitoring (Martin, 1997b).

In the context of this project, the concerns about sustainability expressed by public servants whom I interviewed appear to represent a further questioning of rationalities and techniques of manageable risk, and an assertion of the significance of social forms of governing. For example:

> ...the fact is that there are a range of problems which need other kinds of policy intervention [apart from self-reliance]. Environmental problems are only one example...but we have a pretty horrendous environmental record and we are not going to be able to turn the tide without some form of assistance. What we are talking about at the end of the day is a combination of public and private resources. (Male, Branch Director level [2], DPIE)

> There are problems with the short-term here aren't there, in the sense that self-reliance...like we've got now will mean that farmers will tighten their belts, they won't be able to take the more sensitive conservation-type activities, they'll tend to mine their capital, and there's a lot of evidence that farmers are mining their capital right now and not investing the amounts that they should do to keep their properties going at good productivity, and over-using the land, so, in order to get maximum short-term money.

> The other side of the coin is providing farmers with support like interest subsidies and so forth that may have helped to overstock that land. So there's a two-way type thing. Self-reliance I don't think can address all those economic, environmental and social problems. (Male, Branch Director level [3], DPIE)

> ...self-reliance...can certainly mean that we are setting up a pretty dangerous environment for effective land management. So we have to be a bit careful how we use the terminology and how far we push these philosophies. We don't want to have a big brother approach and the government standing over people. But government representation...needs to reflect the interests of the whole community, and the interest of the community is that the land will be well managed. (Male, Senior Officer level, Qld. Office of Rural Communities)

The previous statements represent the classic liberal principle: one must always suspect that one governs too much (see Foucault, 1997). In other words, as Foucault (1997) notes, governmentality 'should inquire not just to the best (or least costly) means of achieving its effects but also concerning the possibility and even lawfulness of its scheme for achieving effects (74). Thus, assistance for environmental measures is argued here to be good in that it has the potential to optimize the 'natural' regulations of producers, and achieve improved results. However, this results in a tension between the will to govern in an advanced liberal way, and the recognition that environmental issues may require social forms of governance. That is to say, environmental sustainability provides a space through which managed risk is re-ordered in a distinctly social way. This is consistent with both Dean (1997; 1999) and Stenson and Watt (1999) who argue that governance through 'the social' remains an important domain of knowledge that is re-formulated rather than displaced and exists in hybrid forms with advanced liberal strategies of governing.

Representatives from farm organizations also recognize similar limits to self-reliance. However, their arguments reflect the conflict between values of independence and the desire for some form of government intervention – an issue discussed in more detail in Chapter Four.

> Well I mean if you take it [self-reliance] to extremes, I suppose it will [address farm problems], but you would have awful social problems and probably environmental problems and to a lesser extent economic ones. So I think the answer is probably is that yes it can, but the reality is that it will create more problems than it resolves. So there needs to be a balance and a reasonable approach... I mean if you let the marketplace take over there would be very considerable problems. (Male, Executive Member, UGA)

> ...I think there is a whole scope for, certainly for government assistance and intervention in many issues for agriculture such as environmental conservation practices... It has got to be realized that those sorts of things should be borne across the community, whether that be through tax concessions for those sorts of measures or a favourable taxation system for people involved in those sorts of conservation measures. (Male, Executive Member [1], CUA)

> We are not interested in keeping people on the land who aren't viable in the long term. But I think the worst case scenario is to move into a corporate farming system across Australia and there have been demonstrations time and time again that leads to environmental degradation on a much greater scale than you would see with a family farming operation. (Male, Executive Member [2], CUA)

Within these statements there is, again, a desire to optimize the activity of governing. Here, regulation is here seen as necessary, but in a way that supports both the independence of farmers, and the continuation of the family farm. A discourse of environmental sustainability is deployed to assert the necessity of government involvement in ensuring that producers maintain their independence, but in a way that does not threaten the collective livelihood of family farming operations.

A productive tension is thereby created between the desire for farmer autonomy, and the claimed necessity of social forms of support to maintain this independence in a manner that serves the collective interests of all family farmers. As noted in Chapter Four, this ambivalence is shared by producer organizations such as the UGA and the CUA whose constituency is the family farmer/grazier. The NFF – as an umbrella organization of other organisations (and not individual growers) – has a much broader constituency and less of a commitment to small commodity producers (Lawrence, 1987). Therefore, the NFF have tended to take a less favorable approach to social forms of assistance, and to equate sustainability with the amalgamation of farms and exit of inefficient producers (see NFF, 1993). Nevertheless, the rationalities of UGA and CUA representatives, while perhaps not particularly revolutionary, are highly significant in opening a space for the shaping of managed risk in a social way. In other words, the claim that *public* forms of support are necessary to maintain both the optimum functioning, and independence from government, of family farming provides at least some scope for a more critical engagement with management practices presently constructed as truthful.

Conclusion

The three examples of exceptional circumstances, local knowledge and sustainability presented in this chapter suggest that self-reliance as part of an advanced liberal strategy for managing risk has limits imposed on its operation at both a formal political level and among producers themselves. Rather than representing overt resistance to a hegemonic discourse of managed risk, the examples show a more subtle politicization of self-reliance based on social rationalities of governing. In the case of drought exceptional circumstances and sustainability, managed risk was constituted as having limits, while the example of local farmer knowledges demonstrated the differences in rationalities of farm management between state agencies and Central Queensland producers. The limits of managed risk were constructed in terms of distinctly social rationalities of governing in which the collective interests of either farmers or society were seen to be at stake and that demanded public forms of intervention. While perhaps not appearing to represent a challenge to advanced liberal rationalities of governing, this is significant in calling into question the representation of farming as a business enterprize and in providing a site of struggle in which the governing of agricultural change can be debated, politicized and change occur. It can be argued to contribute to the dynamic and contested nature of governing itself in which rule seems to be a constantly 'failing operation' (Miller and Rose, 1990:10). This chapter has argued that the (re)constitution of farm viability in social terms, ensures that strategies seeking to achieve managed risk on the part of farmers face limits, and therefore their effects portend an open-ended set of possibilities.

The final chapter of this book ties together the key themes presented in each of the chapters and draws five conclusions regarding the governance of agricultural change in Australia. It demonstrates in particular the analytical significance of a genealogy of governing and how this approach has the capacity to question the claimed unity of 'projects' of change and make visible the political effects of seemingly neutral forms of knowledge.

Chapter 6

AGRICULTURAL REGULATION: FROM 'PROJECTS' TO 'PROBLEMATIZATIONS' OF RESTRUCTURING

This chapter draws together the arguments presented in the book and makes some conclusions concerning the specific problematizations of governing agricultural change in Australia that have enabled particular rationalities, technologies and forms of expertise to be assembled. In doing so, the conceptual coherence of a genealogical approach to analysing the governance of change is also assessed. Finally, the chapter examines the implications of the conclusions for future research in both agricultural regulation and, more generally, in the field of governmentality studies.

RECONSTRUCTING PROBLEMATIZATIONS OF AGRICULTURAL CHANGE

This book investigated, genealogically, the different ways in which agricultural change in Australia was, from 1967 to 1997, deemed a governable problem. In studying the governing of agricultural change in this way, my aim was to question those *meta-narratives* based upon a search for the origins or foundations of governing. These meta-narratives included primarily economic accounts that viewed farm problems as clearly definable issues caused by claimed normal processes of capitalist development, and which needed to be managed by authorities to ensure an efficient allocation of resources (eg. Kingma and Samuel, 1977; Musgrave, 1990; Gow, 1996). Such accounts have, since the 1960s, been influential in enabling the emergence of a discourse of agricultural restructuring in Australia. Through calling into question the foundations upon which economic explanations of agricultural regulation are constituted as 'truthful', I have in this book contributed to a sociological understanding of this issue.

In making the governing of agricultural change intelligible as a sociological problem, I have contributed also to a critique of structural Marxist approaches that attempt to explain changes in governing with reference to an underlying dynamic or logic. Much of the sociological literature on regulation, and more specifically agricultural regulation, is conducted from a structural Marxist perspective. While these accounts are more critical than economic explanations, and recognize the broader relations of capitalist production within which agricultural regulation is embedded, they tend to unify the governance of agricultural problems according to the claimed imperatives of global capital accumulation. From this perspective, low farm incomes and the exit of farmers are argued to be a logical outcome of global pressures for restructuring that authorities are compelled to address in order to maintain national efficiency and productivity.

In this study I challenged economic understandings of restructuring, and showed the limitations of political economic accounts of regulation, by drawing attention to the complexities and contingencies on which present forms of governing were based. Thus, rather than conceptualizing regulation in terms of distinct projects or epochs of rule driven by foundational dynamics such as economic efficiency or global restructuring, I have argued for an understanding of the *problematizations* that make such 'projects' technically knowable and governable. This was pursued through an exploration of four key issues: 1) the historically specific rationalities and technologies shaping agricultural restructuring as a problem of governing; 2) those vested with the authority to speak the truth about this problem; 3) the types of practices authorized as a result; and, 4) the contestation and re-shaping of these practices. A concern with the different ways in which agricultural change was made a knowable and governable problem disposed this study to a Foucauldian-inspired genealogy that treated knowledge as a perspective. This enabled those meta-narratives, seeking to order history 'in the name of some true knowledge and some arbitrary idea of what constitutes a science and its objects' (Foucault, 1980:83), to be challenged. It allowed also an 'historical knowledge of struggles' (1980:83) to be established to locate these explanations as part of struggles and subjections over power and truth.

An analytics of government provided the theoretical framework through which these ordering attempts could be studied and rendered intelligible. This approach provided neither a theory of how governing 'really' operated, nor claimed to find the key dynamic or actors underpinning strategies seeking to govern restructuring. Rather, it examined the sites, authorities, spaces, forms of thought, and material techniques through which specific problematizations of governing were constituted and shaped in a knowable and manageable form. A 'diagnostic' genealogical methodology enabled restructuring, as a governable problem, to be reconstructed as a series of singular *events* without recourse to a unifying signifier or cause. This approach – a *history of the present* – involved a focus on discursive practices in the assemblage of various ways of reflecting on and managing agricultural change as a governable object. The forms of evidence used to construct a history of the present included archival documents, semi-structured interviews and secondary data. These were useful in providing access to both the contemporary discursive practices of various authorities and the 'local' knowledges of farmers, and in elucidating the historical emergence, from heterogeneous elements, of particular forms of knowledge through which strategies for governing were constructed as truthful.

KEY FINDINGS

Through a genealogical approach to history, the central conclusion of this book is that the now taken-for-granted 'need' for restructuring the managerial capacities of farmers in the claimed interests of economic efficiency is a complex historical product based upon different assemblages of governmental practices. The genealogy of governance applied in each chapter draws attention to the conditions of possibility through which agricultural change has been variously constituted as a 'low-income', 'economic management', 'international competitiveness' or 'individual business acumen' problem. From this perspective, the claimed 'scientific' foundations of governing this problem begin to dissipate into diverse forms of reflecting on and shaping conduct, such as statements on economic management, statistics on farm debt, debates over the managerial capacities of farmers, techniques used to make State authorities more accountable, and the types of droughts considered to be exceptional.

In applying a diagnostic genealogy of government, this book shows that agricultural regulation is more complex than simply 'smoothing', or making more efficient, inevitable processes of change – as economists commonly argue. Equally, it cannot simply be reduced to 'structural' causes of change in which state agencies, in accordance with the requirements of transnational capital, reorganize their agricultural programs in the interests of freer trade and international competition. Such change, and responses to it, are constituted and rendered knowable through a variety of rationalities and technologies of governing. Five key findings drawn from the evidence presented in this book provide evidence for such a conclusion and demonstrate the conceptual coherence of a genealogy of government.

Economic Expertise and the 'Production' of a Restructuring Problem

Present forms of governing agricultural change have emerged from the discursive practices of economic expertise, which played a central role in 'producing' a low-income farm problem from the late-1960s. This book argues that present strategies for governing farmers as prudent business managers can be traced back to the problems identified by economists with expansionist policies in the late-1960s, and attempts to diagnose solutions. The authority attributed to economic explanations of change enabled a questioning of previous forms of governing, and a unified problem of restructuring to emerge out of heterogeneous events. As Chapter One argued, McKay's (1967) paper, in particular, constituted a crucial shift in thinking about how the practices of farmers should best be governed. The questioning of welfarist technologies of governing established through McKay's paper was supported by the later work of Standen and Musgrave (1969) and subsequently by the Head of the Federal Bureau of Agricultural Economics (BAE), Stuart Harris.

Rather than simply describing change, these reflections on previous forms of governing actually constituted new ways of thinking about, categorizing and shaping the practices of farmers. Previous techniques of intervention were based on welfarist forms of support – such as subsidization, tariff support and low-interest credit. Economic expertise, although not accorded the authority it would receive from the early 1970s, was central in the emergence of low-income farms as a knowable restructuring issue. Through the drawing together of heterogeneous elements – such as statistics on farm debt levels and income, commodity prices, cost of production data, and theories of agricultural development – a distinct low-income problem was brought into a calculable domain and constituted as truthful. My research shows that the BAE, as an institutionalized agent of economic expertise, played an important role in constituting previous forms of intervention as widening income disparities between the farm and non-farm sectors.

However, this project found that in focusing primarily on income parity, strategies of governing remained animated by a will to govern through a welfarist rationality of rule, and thereby maintain the economic security of both farmers and society. The greater role given to the BAE from 1967 by the Country Party Minister for Primary Industry, Doug Anthony enabled, via various documents, for a stable material representation of low-income farms to emerge, which could be made amenable to 'evaluation, calculation and intervention' (see Miller and Rose, 1990:7). This was manifested in the creation of 'reconstruction' programs, aimed toward reforming farming industries in the interests of 'distributing the available resources as widely as practicable' (Commonwealth of Australia, 1971a:Schedule, Part I(b)), and in restoring farm incomes to economically 'viable' levels. In responding to the low-income problem, programs of reconstruction – such as the *Rural Reconstruction Scheme* of 1971 – established particular categories of farms and farmers through which assistance could be administered. This was crucial in seeking to differentiate between viable and unviable producers. Programs of farm build-up, debt adjustment and rehabilitation represented the specific technological means for categorizing farmers. These programs sought to ensure that those producers deemed to have sound long-term financial prospects remained in the industry, while those with high debt and low incomes were assisted in 'alleviating conditions of personal hardship' (Commonwealth of Australia, 1971a:Schedule, Part IV) through exiting the sector and finding work elsewhere. While the rationalities underpinning the governing of agricultural change and restructuring have shifted since the early 1970s, economic expertise has remained of central significance in how assistance is administered, and in how the conduct of farmers is reflected upon as a governable problem.

Agricultural Restructuring as an Economic Management Problem

The present focus on agricultural change as a problem of economic management emerged primarily through the creation of the Industries Assistance Commission (IAC) in 1974, and the National Farmers' Federation (NFF) in 1979. The shift, during the latter part of the 1980s and early 1990s, toward a rationality of governing based on the managerial capacities of farmers (see Chapter Three) was underpinned by a belief that assistance to farmers represented a potential threat to overall economic management if not governed in a 'rational' manner. This belief formed an important, yet largely uncontroversial, context in shaping the reforms that took place at the time. However, as the book argues, the evaluation of assistance to farmers based on notions of 'rational' economic management emerged as truthful only in the early 1970s.

Central to the emergence of this way of reflecting on rule was the establishment in 1974 of the IAC, whose task was to evaluate assistance to various industries in terms of their effects on economic efficiency. As argued in Chapter Two, the creation of this statutory body enabled formal technologies of calculation to assume prominence as a more economically authoritative means of evaluating policies of assistance. It also accorded economists authority as experts in using these technologies to evaluate the effect of agricultural policies on the economy. Hence, the authority of economic expertise was reinforced as an authoritative form of knowledge in the preparation of 'professional' submissions concerning assistance levels. This served to de-legitimize the previous practice of farm organizations, many of which had been accustomed to consulting directly with Country Party Ministers over the form and distribution of assistance levels.

The research found that the creation of the IAC undermined welfarist rationalities of governing that had sought to maintain the economic welfare of specific industries. Assistance to these industries was now regarded as a serious issue requiring formal economic evaluation in achieving optimal functioning of a national economy. This was reflected in the IAC's Review in 1976 of the Rural Reconstruction Scheme. The IAC constituted agricultural change as a *structural* problem requiring assistance measures that enabled long-term adjustment to market realities. A focus on the efficiency of farmers as an element of broader national economic management remained a key objective for governing agricultural change until the mid-1980s.

Following its formation in 1979, the NFF favoured an enhanced emphasis on farm adjustment in the interests of greater competition and improved productivity. The NFF was able to moderate the 'sectional' concerns of its affiliated organizations and establish itself as a centre of calculation. As a mouthpiece of large woolgrowers, the NFF opposed so-called protectionist forms of intervention, and favoured the exit of those farmers not considered to have long-term economic prospects. The NFF drew extensively on economic expertise to support its arguments, which were later embraced by the Labor Party, following its election to government in 1983.

While having first assumed an important role through the Federal Bureau of Agricultural Economics (BAE), the creation of the IAC and NFF enabled the techniques and calculations of economists to be given added formal authority in the evaluation of assistance to farmers. I have found that this was of crucial significance in questioning welfarist rationalities of rule that focused on the protection of individual industries, and in linking the viability of farmers to broader national economic management.

Farmers' Managerial Capacities as the Legitimate Site of Reform

The contemporary focus on improving the managerial capacities of farmers is inseparable from the construction of a crisis in national economic management from the mid-1980s, and attempts to 'internationalize' the economy. A focus by authorities during the 1990s on improving farmers' risk management capacities is constituted, in the documents examined for this research, as a logical response to improving the competitive position of Australian agriculture in a sustainable manner. In applying a genealogical approach, this book on the one hand questions such a 'logic', and argues that the emphasis on managerial skills as the 'proper' site of regulation is not the culmination of a progressive shift toward more effective forms of governing change. On the other hand, neither does it imply necessarily a passive response by the state to global pressures for restructuring. In fact, managerial skills were rendered a governable problem through the construction of an economic crisis from 1986, and a belief that improving the capacities of farmers would help resolve this crisis.

The questioning of economic regulation in 1986 based on Treasurer, Paul Keating's 'Banana Republic' statement (Chapter Three), was crucial in shifting the terrain of governing, by constituting a model of economy in which individual farm business capacities could emerge as a problem in achieving improved international competitiveness. This involved the constitution, by a number of authorities, of past economic management techniques as inhibiting 'outward looking' attitudes. This book argues that the NFF played a particularly prominent role in the assembling of an advanced liberal rationality of governing by positioning itself as 'the voice of rural Australia', and questioning 'protectionist' forms of governing favored by the National (Country) Party. As this project identifies, with the support of the NFF improved business practices were promoted by the Labor government – through such documents as *Primary Industries and Resources: Policies for Growth* (Kerin and Cook, 1988), and the 1988 *Rural Adjustment Scheme* (Commonwealth of Australia, 1988a) – as the only way of improving farmers' capacities to compete effectively in international markets.

From the perspective of an analytics of government, a construction of crisis enabled the attitudes and conduct of farmers to assume prominence as the legitimate sites of reform. The 1989 Rural Adjustment Scheme represented the material manifestation of this type of rationality where business management skills were constituted as central in restructuring farmers' conduct, and in defining the boundaries between productive and non-productive producers. A significant part of this shift was the calling into question of farming as a lifestyle, and the belief that a cultural attachment to farming was incompatible with business objectives. The incompatibility of 'welfare' and 'business' objectives is now taken-for-granted in Federal agricultural policy yet, as argued in this book, it has only a very short history.

Self-Reliance as a Strategy for Regulating Managerial Conduct

Self-reliance, as a contemporary strategy for governing agricultural change, represents more than a withdrawal of state intervention. It constitutes an advanced liberal means of governing the capacities of farmers as either 'active citizens' or 'targeted populations'. Contemporary approaches to governing change emphasize the need for farmers to become 'self-reliant' in order to manage in a prudent manner both financial and environmental risk. This view is shared by economists, politicians and the peak farming body, the National Farmers' Federation (NFF). I have argued, however, that self-reliance is more complex than the withdrawal of social forms of support, and their replacement with less direct market-based forms of intervention. It comprises a complex assemblage of strategies and agents for shaping the capacities of farmers in an advanced liberal way. These strategies seek to constitute productivity and environmental sustainability as individual farm management problems that can be addressed through the enhancement of farmers' business skills. Farmers are defined as economically rational actors who, if equipped with 'entrepreneurial' attitudes and formal economic skills, have an improved ability to 'properly' manage their properties, and thus maintain their viability.

Moreover, self-reliance is not the realization of a unified economic rationalist or neo-liberal ideology. A self-reliant approach to governing farm viability has been assembled from a heterogeneous array of elements as outlined in Chapter Four: the reconstitution of self-reliance in terms of rationalities of business management; the construction of sustainability as an individual risk; the questioning of drought as a disaster in 1990, a discursive framework of competitiveness and efficiency, and the belief by both Labor and Liberal/National Coalition governments that 'protectionist' forms of intervention were outmoded. Dean's (1997, 1999) distinction between 'active citizens' and 'targeted populations' provides a conceptually coherent means of rendering intelligible how self-reliance, as an advanced liberal form of governing, seeks to distinguish between, and govern, viable and unviable producers. The promotion of formal managerial skills represents an attempt to define the types of capacities deemed necessary to manage a farm in a profitable and ecologically sustainable manner. Formal training, on the one hand, thus comprises a central technological means of transforming farmers into 'active' managers of risk. Programs such as Property Management Planning (PMP) and FarmBis seek both to instil the 'correct' techniques for prudent conduct into farmers, and establish benchmarks that define how viability is to be best achieved. On the other hand, those producers unable to manage their risk individually are targeted as poor managers who have failed to conduct themselves in an entrepreneurial manner. Such producers are constituted as a burden on the profitability and sustainability of Australian agriculture, and are encouraged to undergo formal counselling so that they come to 'know' that their practices are unviable, and that their future in farming is limited. From this perspective, both 'business' and 'welfare' forms of intervention incorporate technological means for shaping farmers' conduct.

As I argued in the Introduction to this book, a genealogy of governance emphasizes that governing is more complex than the progressive realization of a unified theory, ideology or set of values. The shifts that I have outlined in the various chapters confirm the methodological coherence of this approach and show the ways in which various rationalities and technologies of restructuring have been assembled. Nevertheless, it is insufficient to account for how these assemblages of governing were 'eternally optimistic' in their objectives, yet 'congenitally failing' in their effects (Miller and Rose, 1990). Attention, also needs to be paid also to the agents and forms of knowledge that have sought to shape the limits of rule.

The Limits of Governing Agricultural Change through Self-Reliance

Self-reliance is not indicative simply of the gradual displacement of an interventionist 'welfare-state' by a non-interventionist neo-liberal state. As a domain of knowledge, collectivist rationalities of rule remain of continuing significance in shaping, and setting limits on, the governing of agricultural change. A key finding of this study is that while a problem of agricultural restructuring emerged through the calling into question of welfarist technologies of governing from the 1960s, this did not mean a progressive demise of rationalities seeking to govern through the social. In fact, social forms of governing have been important in defining the limits of strategies seeking to govern in an advanced liberal way. This shows the advantages of a genealogical approach in recognizing the complexities of governance processes, and in drawing attention to those knowledges overlooked by conventional histories in their search for underlying meaning. The significance of social forms of governing is evident in two ways. First, in the forms of contestation to 'managed risk' noted in Chapter Five. Second, in the constant attempts by Commonwealth agencies to achieve, in the absence of direct coercion, conformity by State authorities with national objectives. These tensions have been notable throughout the period under study, and have shaped significantly the limits of agricultural change as a governable issue.

This book demonstrates that the emergence of an advanced liberal regime of governing has not led to the neutralization of those rationalities seeking to govern through the social. In fact, the continuing 'failure' of various programs aimed at enhancing the capacities of farmers to become viable has politicized the strategies involved in achieving improved farm management. Three key sites were examined in this thesis where 'the social' assumes prominence in imposing limits on advanced liberal rationalities of managed risk. Although not exhaustive, these comprise programs of exceptional circumstances, the local knowledge of farmers, and discourses of sustainability.

Chapter Five provides evidence that supports Rose's (1996) claim of a re-configuration of the social. The spaces of contestation examined demonstrate that different forms of rule can exist alongside one another without necessarily compromising rule itself. This finding shows parallels with what Stenson and Watt (1999) call *hybrid* practices of governing and Dean (1999:173) refers to as a post-welfarist regime of the social. While official programs remain animated by a desire to govern in an advanced liberal way, there exist a number of sites where this is rendered problematic, and the conduct of farmers (re)constituted according to welfarist rationalities of governing. Of particular significance is drought as a means of calling into question those rationalities that constitute farmers as individual risk managers. The example of exceptional circumstances and local knowledges showed that drought remains a key site in which the capacities of farmers are shaped in a distinctly social way.

Also important in providing a space for the emergence of hybrid practices of rule are the historical relations between Commonwealth agencies and State authorities. As noted in Chapters One to Four, the States have, since 1971, assumed responsibility for the day-to-day administration of farm reconstruction/adjustment programs. While the regulation of agricultural production is a responsibility of the States, the recognition of problems that extend beyond State borders has compelled the Commonwealth to assume involvement. However, this has been possible only through the administration of grants under section 96 of the Federal Constitution. The division of powers between the Commonwealth and States concerning issues of agricultural change has led to a number of tensions. State governments have a long history of opposition to attempts by the Commonwealth to exercize authority over their affairs. This has meant that Federal-State agreements must be relatively flexible to enable the achievement of 'national' objectives. Nevertheless, Commonwealth objectives have frequently been compromised from the outset. Thus, attempts to introduce greater standardization in State assessments of viability have encountered persistent problems. This represents one of the central issues in governing 'at a distance' (see Miller and Rose, 1990). It indicates that the administration at the State level remains an important means of shaping Commonwealth objectives in a way that emphasizes restructuring as more than simply a problem located in the managerial capacities of individual producers. Despite the introduction of technologies of performance noted in Chapter Three, aimed toward aligning the conduct of State authorities with Commonwealth objectives, it seems that 'social' rationalities of risk management will continue to shape the governing of agricultural change, and define the limits of advanced liberal rule.

Implications for Future Research on Regulation

The research on which this book is based commenced in early 1997 during a period of considerable change in agricultural policy and rural politics. Pauline Hanson's right-wing One Nation Party was at the height of its political prominence in contesting the historical National (Country) Party dominance of the rural vote, many parts of Queensland and New South Wales remained in the grip of drought, commodity prices for beef were very low, there was concern from many farmers about the adverse effects of global competition on their terms-of-trade, the Mid-Term Review of the Rural Adjustment Scheme was released mid-year, and there was uncertainty as to the future direction of agricultural programs and funding. Following the release in late 1997 of the *Agriculture – Advancing Australia* ('Triple A') policy package, however, there emerged a semblance of stability in how agricultural change was reflected upon and governed. This has been despite the enormous organizational changes to the former Department of Primary Industries and Energy (now Agriculture, Forestry and Fisheries Australia). While the spectres of drought, One Nation, low commodity prices and international competition remain, the strategies and programs deployed to govern agricultural change has been retained, and the emphasis on self-reliance and managed risk altered little. Any changes have been primarily in the allocation of funding for specific programs.

Nevertheless, two recent changes to the 'Triple A' warrant attention. The first of these was the introduction of the Farm Innovation Program launched on 9 May 2000. This program forms part of the Triple A package and provides grants of up to 50 per cent of eligible project costs for innovative agricultural processes, products and practices. The second change of note was the announcement in October 2000 that the FarmBis and Property Management Planning (PMP) programs would be amalgamated. This amalgamation represents an ongoing commitment at a Federal government level to practices of business planning and managed risk noted in this book. In many ways, the choice to end the analysis of the governing of change in 1997 has been vindicated. The material presented throughout the book remains of clear significance to present agricultural policy issues and concerns. Continuation of the practices of governing noted in Chapters Four and Five raise some significant issues for future research in farm policy, agricultural regulation and governmentality studies. They point also toward potential sites for further inquiry, the scope of which this book was unable to adequately address.

As Lockie (1999) notes, sociological approaches to agricultural restructuring have long had a focus on searching for the central actor or dynamic driving changes in farm policies. This book demonstrates that this type of analysis – endorsed by both economists and many sociologists – has a number of limitations in accounting for the complexity and contingency of governing. In particular, it is unable to account adequately for the heterogeneous elements through which 'problems' or 'agenda' of restructuring are assembled and made knowable. The governing of agricultural change is argued in this book to have been made possible through political rationalities and technologies that cannot be traced to a foundational dynamic or logic. As a consequence, future studies of restructuring and the regulation of agriculture need to pay more attention to the chance events, political alignments, forms of knowledge, agents, and techniques of shaping conduct that enable particular ways of reflecting on farm practice to emerge as governable issues. Murdoch and Ward (1997) provide a good example of how this type of analysis might be conducted. However, a focus on the specific forms of 'translation' between various agents, such as Commonwealth and State authorities, farm organizations and farmers in rendering agricultural restructuring knowable, calculable and manageable would illuminate further, than this book has been able to do, the complexities and 'messiness' of governing in its practice (see O'Malley et al., 1997; Stenson, 1998; Barry, 2001). In addition, such analysis does not necessarily need to be confined to specific nation-states. The increasing role of global regulatory bodies and multinational agribusiness corporations in agricultural regulation and development raises the question of how these organizations act as agents of governmentality (see Dillon, 1995 and Brigg, 2001 as examples of how this can be achieved). Future research could examine the rationalities and technologies that constitute these seemingly 'global' bodies, how their capacities are defined and contested, and how this shapes the practices of other actors.

In applying an analytics of government to the governing of agricultural change, I argue that more attention needs to be paid in governmentality studies to what Stenson and Watt (1999) call 'hybrid' forms of rule. The evidence examined shows that while distinct 'regimes of governing' (Dean, 1999) are discernable in the governance of agricultural change, these comprise complex and shifting relations between social and advanced liberal rationalities and technologies of rule. Future research needs to examine more closely how 'the social' exists with, and imposes limits on, advanced liberal forms of governing. In particular, further attention ought to be placed on 'local' knowledges (see O'Malley, 1996) – raised only in a preliminary way in this thesis – and the ways in which these shape strategies of governing. I have shown elsewhere the significance of local knowledge in the shaping of 'sustainable' natural resource management strategies (see Higgins et al., 2001; see also Clark and Murdoch, 1997). There is potential to develop further this type of analysis to studies of environmental management to show the mutual shaping of strategies of rule. Following the lead of O'Malley (1996), this could also be applied more generally to studies of governmentality.

The construction of a genealogy of governing in this book shows the significance of how various problematizations of governing agricultural restructuring have been constituted in a thinkable and manageable form, and the heterogeneous associations that made this possible. In addition, it emphasizes the contingent foundations upon which present forms of governing are based. It appears that a genealogical approach to regulation is well placed to question the seemingly stable and certain foundations claimed to characterize present forms of governing.

REFERENCES

ABARE (1997) 'Changing structure of farming', *ABARE Current Issues*, **4**: 1-7.

Aglietta, M. (1987) *A Theory of Capitalist Regulation: The US Experience* (D. Fernbach, trans.), London: Verso.

Aitkin, D. (1988) ''Countrymindedness' – the spread of an idea', in S.L. Goldberg and F.B. Smith (eds), *Australian Cultural History*, Cambridge: Cambridge University Press: 50-57.

Axford, B. (1995) *The Global System: Economics, Politics and Culture*, Cambridge: Polity Press.

Balderstone, J.S, Duthie, L.P., Jarrett, F.G., Eckersley, D.P. and McColl, J.C. (1982) *Agricultural Policy: Issues and Options for the 1980s. Working Group Report to the Minister for Primary Industry*, Canberra: Australian Government Publishing Service.

Bamberry, G., Dunn, T. and Lamont, A. (1997) *A Pilot Study to Investigate the Relationship Between Farmer Education and Good Farm Management*, Canberra: Rural Industries Research and Development Corporation.

Barr, N. and Ridges, S. (1998) *Structural Adjustment and the Implementation of Catchment Management Plans*, Summary of a report prepared for the Murray Darling Basin Commission, Melbourne: Victorian Dept. of Natural Resources and Environment.

Barry, A. (2001) *Political Machines: Governing a Technological Society*, London: The Athlone Press.

Beeson, M. and Firth, A. (1998) 'Neoliberalism as a political rationality: Australian public policy since the 1980s', *Journal of Sociology*, **34**(3): 215-231.

Bird, J.G. and Marriott, B.W. (1966) 'A regional case study of dairy industry adjustment', *Review of Marketing and Agricultural Economics*, **34**(1): 202-211.

Bonanno, A. (1994) 'The locus of polity action in a global setting', in A. Bonanno, L. Busch, W. Friedland, L. Gouveia and E. Mingione (eds), *From Columbus to ConAgra: The Globalization of Agriculture and Food*, Lawrence: University Press of Kansas: 251-264.

Bonanno, A., Busch, L., Friedland, W., Gouveia, L. and Mingione, E. (1994) 'Introduction', in A. Bonanno, L. Busch, W. Friedland, L. Gouveia and E. Mingione (eds), *From Columbus to ConAgra: The Globalization of Agriculture and Food*, Lawrence: University Press of Kansas: 1-26.

Brigg, M. (2001) 'Empowering NGOs: the micro-credit movement through Foucault's notion of *Dispositif*', *Alternatives*, **26**: 233-258.

Bryson, L. (1992) *Welfare and the State: Who Benefits?* London: Macmillan.

Burdon, A. (1996) 'Commonwealth government assistance for adjustment in agriculture', in Dept. of the Parliamentary Library (eds), *Australian Rural Policy Papers 1990-1995*, Canberra: Australian Government Publishing Service: 1-48.

Bureau of Agricultural Economics (1971) *Debt Reconstruction and Farm Adjustment: The Immediate and Longer Term Needs for Debt Reconstruction and Farm Adjustment with Special Reference to the Sheep Industry*, Canberra: Australian Government Publishing Service.

Bureau of Agricultural Economics (1973) *The Rural Reconstruction Scheme: A Review of Progress*, Canberra: Australian Government Publishing Service.

Bureau of Agricultural Economics (1974) *The Rural Reconstruction Scheme: A Review of Progress (Second Report)*, Canberra: Australian Government Publishing Service.

Burns, R. (1994) *Introduction to Research Methods* (2nd edn), Melbourne: Longman.

Butlin, N.G., Barnard, A. and Pincus, J.J. (1982) *Government and Capitalism: Public and Private Choice in Twentieth Century Australia*, Sydney: George Allen and Unwin.

Buttel, F. (1996) 'Theoretical issues in global agri-food restructuring', in D. Burch, R. Rickson and G. Lawrence (eds), *Globalization and Agri-Food Restructuring: Perspectives from the Australasia Region*, Aldershot: Avebury: 17-44.

Callon, M. (1986) 'Some elements of a sociology of translation: domestication of the scallops and fishermen of St Brieuc Bay', in J. Law (ed.), *Power, Action and Belief: A New Sociology of Knowledge?* London: Routledge and Kegan Paul: 196-233.

Callon, M. and Latour, B. (1981) 'Unscrewing the big Leviathan: how actors macro-structure reality and how sociologists help them to do so', in K. Knorr-Cetina and A.V. Cicourel (eds), *Advances in Social Theory and Methodology: Toward an Integration of Micro- and Macro-Sociologies*, Boston: Routledge and Kegan Paul: 277-303.

Campbell, H. and Lawrence, G. (2000) 'Assessing the neoliberal experiment in Antipodean agriculture: an investigation into the 'sociology of instability'', Paper presented at the Xth World Congress of Rural Sociology, *Sustainable Rural Livelihoods: Building Communities, Protecting Resources, Fostering Human Development*, Rio de Janeiro, 20 July-5 August.

Campbell, K.O. (1974) 'Rural policy in Australia – 1974 style', *The Australian Journal of Agricultural Economics*, **18**(3): 157-170.

Campbell, K.O. (1980) *Australian Agriculture: Reconciling Change and Tradition*, Melbourne: Longman Cheshire.

Cassidy, P.A., McCarthy, O. and Hamilton, C.P. (1969) 'Adjustment problems in Queensland agriculture', in J.P. Makeham and J.G. Bird (eds), *Problems of Change in Australian Agriculture*, Armidale: Department of University Extension, University of New England: 161-180.

Catley, B. (1996) *Globalizing Australian Capitalism*, Cambridge: Cambridge University Press.

Chamala, S., Coughenour, C. and Keith, K. (1983) *Study of Conservation Cropping on the Darling Downs – A Basis for Extension Programming*, Brisbane: Dept. of Agriculture, University of Queensland.

Cheers, B. (1998) *Welfare Bushed: Social Care in Rural Australia*, Aldershot, UK: Ashgate.

Clark, J. and Murdoch, J. (1997) 'Local knowledge and the precarious extension of scientific networks: a reflection on three case studies', *Sociologia Ruralis*, **37**(1): 38-60.

Cockfield, G. (1995) *From Closer Settlement to Rural Adjustment: The Theory and Practice of Broadacre Farm Structure Policies in Australia*, Armidale: Unpublished Master of Public Policy Thesis, University of New England.

Commonwealth of Australia (1969) 'Appropriation Bill (No. 1)', *Parliamentary Debates, House of Representatives*, (Vol. 65) Canberra: Australian Government Publishing Service.

Commonwealth of Australia (1970a) *Marginal Dairy Farms Agreement Act*, Canberra: Australian Government Publishing Service.

Commonwealth of Australia (1970b) 'Marginal Dairy Farms Agreement Bill', *Parliamentary Debates, House of Representatives*, (Vols 67/68) Canberra: Australian Government Publishing Service.

Commonwealth of Australia (1970c) 'Australian Rural Economy: Discussion of Matter of Public Importance', *Parliamentary Debates, House of Representatives*, (Vol. 69) Canberra: Australian Government Publishing Service.

Commonwealth of Australia (1971a) *States Grants (Rural Reconstruction) Act*, Canberra: Australian Government Publishing Service.

Commonwealth of Australia (1971b) 'Rural Reconstruction: Discussion of Matter of Public Importance', *Parliamentary Debates, House of Representatives*, (Vol. 72) Canberra: Australian Government Publishing Service.

Commonwealth of Australia (1971c) 'States Grants (Rural Reconstruction) Bill', *Parliamentary Debates, House of Representatives*, (Vol. 72) Canberra: Australian Government Publishing Service.

Commonwealth of Australia (1973a) 'Industries Assistance Commission Bill', *Parliamentary Debates, House of Representatives*, (Vol. 87) Canberra: Australian Government Publishing Service.

Commonwealth of Australia (1973b) *Industries Assistance Commission Act*, Canberra: Australian Government Publishing Service.

Commonwealth of Australia (1976a) *States Grants (Rural Adjustment) Act*, Canberra: Australian Government Publishing Service.

Commonwealth of Australia (1976b) *Parliamentary Debates, House of Representatives*, (Vol. 102) Canberra: Australian Government Publishing Service.

Commonwealth of Australia (1985a) *States and Northern Territory Grants (Rural Adjustment) Act*, Canberra: Australian Government Publishing Service.

Commonwealth of Australia (1985b) *Parliamentary Debates, House of Representatives*, (Vol. 141) Canberra: Australian Government Publishing Service.

Commonwealth of Australia (1986) *Economic and Rural Policy: A Government Policy Statement*, Canberra: Australian Government Publishing Service.

Commonwealth of Australia (1988a) *States and Northern Territory Grants (Rural Adjustment) Act*, Canberra: Australian Government Publishing Service.

Commonwealth of Australia (1988b) *Parliamentary Debates, House of Representatives*, (Vol.163) Canberra: Australian Government Publishing Service.

Commonwealth of Australia (1990) *Managing for Drought: Final Report of the Drought Policy Review Task Force*, Canberra: Australian Government Publishing Service.

Commonwealth of Australia (1992a) *National Drought Policy – Appropriate Government Responses to the Recommendations of the Drought Policy Review Task Force. First*

Report from the Senate Standing Committee on Rural and Regional Affairs, Canberra: Australian Government Publishing Service.

Commonwealth of Australia (1992b) *National Drought Policy: Managing for Risk and Productivity*, Canberra: Department of Primary Industries and Energy.

Commonwealth of Australia (1992c) *Parliamentary Debates, House of Representatives*, (Vol. 186) Canberra: Australian Government Publishing Service.

Commonwealth of Australia (1994) *Rural Adjustment, Rural Debt and Rural Reconstruction: Report of the Senate Rural and Regional Affairs and Transport References Committee*, Canberra: Australian Government Publishing Service.

Commonwealth of Australia (1995) *Managing for the Future. Report of the Land Management Task Force*, Canberra: Australian Government Publishing Service.

Commonwealth of Australia (1996) *Report of the Drought Policy Review Task Force*, Canberra: Australian Government Publishing Service.

Connors, T. (1996) *To Speak with One Voice: The Quest by Australian Farmers for Federal Unity*, Canberra: National Farmers' Federation.

Cook, V., Edwards, K. and Ronan, G. (eds) (1994) *A Positive Approach to Farm Adjustment*, Adelaide: Primary Industries South Australia.

Coopers and Lybrand WD Scott (1988) *Report on the Review of the Rural Adjustment Scheme*, Canberra: Department of Primary Industries and Energy.

Craig, J. (1991) *Australian Politics Notebook*, Sydney: Harcourt, Brace, Jovanovich.

Dean, M. (1994) *Critical and Effective Histories: Foucault's Methods and Historical Sociology*, London: Routledge.

Dean, M. (1996) 'Putting the technological into government', *History of the Human Sciences*, **9**(3): 47-68.

Dean, M. (1997) 'Sociology after society', in D. Owen (ed.), *Sociology after Postmodernism*, London: Sage: 205-228.

Dean, M. (1999) *Governmentality: Power and Rule in Modern Society*, London: Sage.

Dempsey, K. (1990) *Smalltown*, Melbourne: Oxford University Press.

Denzin, N.K. (1978) *The Research Act*, New York: McGraw-Hill.

Department of Primary Industries and Energy (1998a) *National Rural Finance Summit: Report of the Activating Committee – Outline*, URL: http://www.affa.gov.au/rural/conference/fs/report/outline.html.

Department of Primary Industries and Energy (1998b) *National Rural Finance Summit: Report of the Activating Committee – Education and Training,* URL: http://www.affa.gov.au/rural/conference/fs/report/education.html.

Department of Primary Industries and Energy (1998c) *National Rural Finance Summit: Report of the Activating Committee – Adjustment,* URL: http://www.affa.gov.au/rural/conference/fs/report/adjustment.html.

Department of Primary Industries and Energy (1998d) *Agriculture – Advancing Australia*, URL: http://www.dpie.gov.au/aaa.html.

Department of the Prime Minister and Cabinet (1991) *Building a Competitive Australia*, Canberra: Australian Government Publishing Service.

Dillon, M. (1995) 'Sovereignty and governmentality: from the problematics of the 'new world order' to the ethical problematic of world order', *Alternatives*, **20**: 323-368.

Dovers, S. (1992) 'The history of natural resource use in Australia: practicalities and ideologies', in G. Lawrence, F. Vanclay and B. Furze (eds), *Agriculture, Environment and Society: Contemporary Issues for Australia*, Melbourne: Macmillan: 1-18.

Edwards, G.W. and Watson, A.S. (1978) 'Agricultural policy', in F.H. Gruen (ed.), *Surveys of Australian Economics*, Sydney: George Allen and Unwin: 189-239.

Encel, S. (1970) *Equality and Authority: A Study of Class, Status and Power in Australia*, Melbourne: Cheshire.

Fontana, A. and Frey, J.H. (1994) 'Interviewing: the art of science', in N.K. Denzin and Y.S. Lincoln (eds), *Handbook of Qualitative Research*, Thousand Oaks, CA: Sage: 361-376.

Foucault, M. (1972) *The Archaeology of Knowledge* (A.M. Sheridan Smith, trans.), London: Tavistock.

Foucault, M. (1980) 'Two Lectures', in C. Gordon (ed.), *Power/Knowledge: Selected Interviews and Other Writings 1972-77* (C. Gordon, L. Marshall, J. Mepham and K. Soper, trans.), Brighton: Harvester Press: 78-108.

Foucault, M. (1982) 'Afterword: the subject and power', in H. Dreyfus and P. Rabinow (eds), *Michel Foucault: Beyond Structuralism and Hermeneutics*, Sussex: Harvester Press: 208-226.

Foucault, M. (1984a) 'Nietzsche, genealogy, history', in P. Rabinow (ed.), *The Foucault Reader*, New York: Pantheon Books: 76-100.

Foucault, M. (1984b) 'The order of discourse', in M. Shapiro (ed.), *Language and Politics*, New York: New York University Press: 108-138.

Foucault, M. (1988) 'Technologies of the self', in L.H. Martin, H. Gutman and P.H. Hutton (eds), *Technologies of the Self: A Seminar with Michel Foucault*, Amherst: Massachusetts: 16-49.

Foucault, M. (1991) 'Governmentality', in G. Burchell, C. Gordon and P. Miller (eds), *The Foucault Effect: Studies in Governmentality*, London: Harvester Wheatsheaf: 87-104.

Foucault, M. (1997) 'The birth of biopolitics', in P. Rabinow, (ed.), *Ethics: Subjectivity and Power: The Essential Works of Michel Foucault 1954-1984 (Vol. 1)*, London: Penguin: 73-79.

Garnaut, J., Robinson, J. and Lubulwa, M. (1997) *Issues in the Delivery of Commonwealth Social Support Programs to Farm Families*, Canberra: ABARE Research Report 97.7, Australian Bureau of Agricultural and Resource Economics.

Garnaut, J. and Lim-Applegate, H. (1998) *People in Farming*, ABARE Research Report 98.6, Canberra: Australian Bureau of Agricultural and Resource Economics.

Glesne, C. and Peshkin, A. (1992) *Becoming Qualitative Researchers: An Introduction*, New York: Longman.

Gow, J. (1996) 'Structural adjustment in Australian agriculture revisited', *Rural Society*, **6**(1): 24-29.

Gow, J. (1997) 'Commonwealth drought policy: 1989-1995: a case study of economic rationalism', *Australian Journal of Social Issues*, **32**(3): 273-282.

Gow, J. and Stayner, R. (1995) 'The process of farm adjustment: a critical review, *Review of Marketing and Agricultural Economics*, **63**(2): 272-283.

Gray, I. (1991) 'Family farming and ideology: some preliminary exploration', in M. Alston (ed) *Family Farming: Australia and New Zealand. Key Papers No.2*, Wagga Wagga: Centre for Rural Social Research, Charles Sturt University – Riverina: 55-68.

Gray, I. and Lawrence, G. (2001) *A Future for Regional Australia: Escaping Global Misfortune*, Melbourne: Cambridge University Press.

Gray, I., Dunn, T. and Lawrence, G. (1993) *Coping with Change: Australian Farmers in the 1990s*, Wagga Wagga: Centre for Rural Social Research, Charles Sturt University – Riverina.

Graycar, A. (1983) 'Retreat from the welfare state', in A. Graycar (ed.), *Retreat from the Welfare State: Australian Social Policy in the 1980s*, Sydney: George Allen and Unwin: 1-12.

Gruen, F. (1990) 'Economic development and agriculture since 1945', in D.B. Williams (ed.), *Agriculture in the Australian Economy* (3rd edn), Sydney: Sydney University Press: 19-26.

Gruen, F. and Grattan, M. (1993) *Managing Government: Labor's Achievements and Failures*, Melbourne: Longman Cheshire.

Hardaker, J.B., Lewis, J.N. and McFarlane, G.C. (1970) *Farm Management and Agricultural Economics: An Introduction*, Sydney: Angus and Robertson.

Harris, S.F. (1970) 'Farm adjustment and the role of government', *The Australian Journal of Agricultural Economics*, **14**(2): 107-120.

Harris, S., Crawford, J., Gruen, F. and Honan, N. (1974) *Rural Policy in Australia. A Working Group Report*, Canberra: Australian Government Publishing Service.

Head, B.W. (1989) Parties and the Policy Agenda 1978-1988, in B.W. Head and A. Patience (eds), *From Fraser to Hawke: Australian Public Policy in the 1980s*, Melbourne: Longman Cheshire: 486-510.

Heady, E.O. (1966) *Agricultural Problems and Policies in Developed Countries*, Oslo: Bondenes Forlag.

Hefford, R.K. (1985) *Farm Policy in Australia*, Brisbane: University of Queensland Press.

Held, D. (1991) 'Democracy, the nation-state and the global system', *Economy and Society*, **20**(2): 138-172.

Henderson, R.F., Harcourt, A. and Harper, R.J.A. (1970) *People in Poverty*, Melbourne: Cheshire.

Higgins, V. (1999) 'Economic restructuring and neo-liberalism in Australian rural adjustment policy' in D. Burch, J. Goss and G. Lawrence (eds), *Restructuring Global and Regional Agricultures: Transformations in Australasian Agri-food Economies and Spaces*, Aldershot: Ashgate: 131-143.

Higgins, V., Lockie, S. and Lawrence, G. (2001) 'Governance, 'local' knowledge and the adoption of sustainable farming practices', in G. Lawrence, V. Higgins and S. Lockie (eds), *Environment, Society and Natural Resource Management: Theoretical Perspectives from Australasia and the Americas*, Cheltenham, UK: Edward Elgar: 212-224.

Hindess, B. (1998) 'Neo-liberalism and the national economy', in M. Dean and B. Hindess (eds), *Governing Australia: Studies in Contemporary Rationalities of Government*, Cambridge: Cambridge University Press: 210-226.

Hirst, P. and Thompson, G. (1996) *Globalization in Question: The International Economy and the Possibilities of Governance*, Cambridge: Polity Press.

Hyde, C.E. (1997) *The Rural Adjustment Scheme: An Economic Analysis*, Melbourne: Dept. of Economics, University of Melbourne.

Industries Assistance Commission (1976) *Rural Reconstruction*, Canberra: Australian Government Publishing Service.

Industries Assistance Commission (1983) *Rural Adjustment (Interim Report)*, Canberra: Australian Government Publishing Service.

Industries Assistance Commission (1984) *Rural Adjustment*, Canberra: Australian Government Publishing Service.

Industry Commission (1996) *Industry Commission Submission to the Mid-Term Review of the Rural Adjustment Scheme*, Canberra: Australian Government Publishing Service.

Jaensch, D. (1992) *The Politics of Australia*, Melbourne: Macmillan.

Jarrett, F.G. (1983) 'The Balderstone Report: an overview', *Australian Journal of Agricultural Economics*, **27**(2): 117-123.

Jessop, B. (1994a) 'Post-fordism and the state', in A. Amin (ed.), *Post-Fordism: A Reader*, Oxford: Blackwell: 251-279.

Jessop, B. (1994b) 'The transition to post-Fordism and the Schumpeterian workfare state', in R. Burrows and B. Loader (eds), *Towards a Post-Fordist Welfare State*, London: Routledge: 13-37.

Jessop, B. (1997) 'Capitalism and its future: remarks on regulation, government and governance', *Review of International Political Economy*, **4**(3): 561-581.

Kapferer, J. (1990) 'Ruralism and nationalist ideology', *Australia and New Zealand Journal of Sociology*, **26**(1): 87-106.

Keating, P. (1988) *May Economic Statement*, Canberra: Australian Government Publishing Service.

Kelly, P. (1992) *The End of Certainty: The Story of the 1980s*, Sydney: Allen and Unwin.

Kerin, J. (1983) 'Rural and regional policy', in J. Reeves and K. Thompson (eds*), Labor Essays 1983*, Melbourne: Drummond: 54-66.

Kerin, J. and Cook, P. (1988) *Primary Industries and Resources: Policies for Growth*, Canberra: Australian Government Publishing Service.

Keynes, J.M. (1936) *The General Theory of Employment, Interest and Money*, London: Macmillan.

Kidman, M. (1991) 'New town: belonging, believing and bearing down', in M. Alston (ed) *Family Farming: Australia and New Zealand. Key Papers No.2*, Wagga Wagga: Centre for Rural Social Research, Charles Sturt University – Riverina: 35-53.

Kilpatrick, S. (1996) *Change, Training and Farm Profitability*, NFF Research Papers, Vol. 10, Canberra: National Farmers' Federation.

Kilpatrick, S., Johns, S., Murray-Prior, R. and Hart, D. (1999) *Managing Farming: How Farmers Learn (Final Report)*, Canberra: Rural Industries Research and Development Corporation.

Kingma, O. and Samuel, S. (1977) 'An economic perspective of structural adjustment in the rural sector', *Quarterly Review of Agricultural Economics*, **30**(3): 201-215.

Kloppenburg, J. (1991) 'Social theory and the de/reconstruction of agricultural science: local knowledge for an alternative agriculture', *Rural Sociology*, **56**(4): 519-548.

Latour, B. (1988) 'Drawing things together', in M. Lynch and S. Woolgar (eds), *Representation in Scientific Practice*, Cambridge, MA: MIT Press: 19-68.

Lawrence, G. (1987) *Capitalism and the Countryside: The Rural Crisis in Australia*, Sydney: Pluto Press.

Lawrence, G. (1992) 'Farm structural adjustment: the imperative for the nineties?', *Rural Society*, **2**(4): 5-7.

Lawrence, G. (1995) *Futures for Rural Australia: From Agricultural Productivism to Community Sustainability*, Rockhampton: Rural Social and Economic Research Centre, Central Queensland University.

Lawrence, G. (1996) 'Rural Australia: insights and issues from contemporary political economy', in G. Lawrence, K. Lyons and S. Momtaz (eds), *Social Change in Rural Australia*, Rockhampton: Rural Social and Economic Research Centre, Central Queensland University: 332-349.

Lawrence, G. and Vanclay, F. (1994) 'Agricultural change in the semiperiphery: the Murray-Darling Basin, Australia', in P. McMichael (ed.), *The Global Restructuring of Agro-Food Systems*, Ithaca, NY: Cornell University Press: 76-103.

Le Heron, R. (1993) *Globalised Agriculture, Political Choice*, Oxford: Pergamon Press.

Lembit, M. and Kingma, O. (1995) 'Commonwealth drought policies', Paper presented at *Coping with Drought: The Australian Institute of Agricultural Science Drought Forum*, Sydney, April 6-7.

Lewis, J.N. (1961) 'Agricultural adjustment in a changing institutional setting', *Journal of the Australian Institute of Agricultural Science*, **27**(4): 213-219.

Lewis, J.N. (1969) 'Problems of institutional change in Australian agriculture', in J.P. Makeham and J.G. Bird (eds), *Problems of Change in Australian Agriculture*, Armidale: Department of University Extension, University of New England: 225-232.

Lincoln, Y.S. and Guba, E.G. (1985) *Naturalistic Inquiry*, London: Sage.

Lipietz, A. (1987) *Mirages and Miracles*, London: New Left Books.

Lipietz, A. (1992) *Towards a New Economic Order. Postfordism, Ecology and Democracy*, New York: Oxford University Press.

Lockie, S. (1999) 'The state, rural environments, and globalisation: 'action at a distance' via the Australian Landcare program', *Environment and Planning A*, **31**: 597-611.

Long, N. and Long, A. (eds) (1992) *Battlefields of Knowledge: The Interlocking of Theory and Practice in Social Research and Development*, London: Routledge.

Long, N. and Villareal, M. (1994) 'The interweaving of knowledge and power in development interfaces', in I. Scoones and J. Thompson (eds), *Beyond Farmer First: Rural People's Knowledge, Agricultural Research and Extension Practice*, London: Intermediate Technology Publications: 41-52.

Lyotard, J. (1984) *The Postmodern Condition: A Report on Knowledge* (G. Bennington and B. Massumi, trans.), Manchester: Manchester University Press.

Makeham, J.P. and Bird, J.G. (eds) (1969) *Problems of Change in Australian Agriculture*, Department of University Extension, University of New England, Armidale.

Malcolm, B., Davidson, B. and Vandenburg, E. (2000) *The Rural Adjustment Scheme: Its Role, Operation and Effectiveness*, Canberra: Rural Industries Research and Development Corporation.

Malcolm, L.R. (1989) 'Rural industry policies', in B.W. Head and A. Patience (eds), *From Fraser to Hawke: Australian Public Policy in the 1980s*, Melbourne: Longman Cheshire: 132-158.

Marsden, T. (1992) 'Exploring a rural sociology for the Fordist transition: incorporating social relations into economic restructuring', *Sociologia Ruralis*, **32**(2): 209-230.

Marshall, N. (1985) 'Rural interest groups', in B. Costar and D. Woodward (eds), *Country to National: Australian Rural Politics and Beyond*, Sydney: George Allen and Unwin: 23-34.

Martin, W. (1990) 'Rural policy', in Jennet, C. and Stewart, R.G. (eds), *Hawke and Australian Public Policy: Consensus and Restructuring*, Melbourne: Macmillan: 155-179.

Martin, P. (1997a) 'The constitution of power in Landcare: a post-structuralist perspective with modernist undertones', in S. Lockie and F. Vanclay (eds), *Critical Landcare*, Key Papers No. 5, Wagga Wagga: Centre for Rural Social Research, Charles Sturt University – Riverina: 45-56.

Martin, P. (1997b) 'Saline politics: local participation and neoliberalism in Australian rural environments', *Space and Polity*, **1**(1): 115-133.

McCarthy, W.O., Hodgson, H. and Briggs, J.E. (1966) 'Dairy farm amalgamation and increased farm output', *Review of Marketing and Agricultural Economics*, **34**: 29-35.

McColl, J., Donald, R. and Shearer, C. (1997) *Rural Adjustment: Managing Change - Mid-term Review of the Rural Adjustment Scheme*, Canberra: Department of Primary Industries and Energy.

McKay, D.H. (1967) 'The small-farm problem in Australia', *The Australian Journal of Agricultural Economics*, **11**(2): 115-132.

McKenzie, D. (1997a) 'Farmers count cost of reform as loan aid cut', *The Australian*, May 14: 23.

McKenzie, D. (1997b) 'Dignified exits for desperate farmers', *The Weekend Australian*, May 17-18: 6.

McMichael, P. (1996) *Development and Social Change: A Global Perspective*, Thousand Oaks, CA: Pine Forge Press.

McMichael, P. (1999) 'Virtual capitalism and agri-food restructuring', in D. Burch, J. Goss and G. Lawrence (eds), *Restructuring Global and Regional Agricultures: Transformations in Australian Agri-Food Economies and Spaces*, Aldershot: Ashgate: 3-22.

McMichael, P. and Myhre, D. (1991) 'Global regulation vs. the nation state: agro-food systems and the new politics of capital', *Capital and Class*, **43**(2): 83-106.

Merriam, S.B. (1998) *Qualitative Research and Case Study Applications in Education*, San Francisco: Jossey-Bass.

Millar, J. and Curtis, A. (1999) 'The nature and role of farmer knowledge in temperate pasture management in the Murray-Darling Basin', *Rural Society*, 9(1): 301-312.

Miller, P. and Rose, N. (1990) 'Governing economic life', *Economy and Society*, **19**(1): 1-31.

Minichiello, V., Aroni, R., Timewell, E. and Alexander, L. (1990) *In-Depth Interviewing: Researching people*, Melbourne: Longman Cheshire.

Mishra, R. (1984) *The Welfare State in Crisis: Social Thought and Social Change*, Brighton: Wheatsheaf.

Muetzelfeldt, M. (1992) 'Economic rationalism in its social context', in M. Muetzelfeldt (ed.), *Society, State and Politics in Australia*, Sydney: Pluto Press: 187-215.

Murdoch, J. (1995) 'Governmentality and the politics of resistance in UK agriculture: the case of the Farmers' Union in Wales', *Sociologia Ruralis*, **35**(2): 187-205.

Murdoch, J. and Clark, J. (1994) 'Sustainable knowledge', *Geoforum*, **25**(2): 115-132.

Murdoch, J. and Ward, N. (1997) 'Governmentality and territoriality: the statistical manufacture of Britain's 'national farm'', *Political Geography*, **16**(4): 307-324.

Musgrave, W.F. (1990) 'Rural adjustment', in D.B. Williams (ed.), *Agriculture in the Australian Economy*, Sydney: Sydney University Press: 248-265.

National Farmers' Federation (1981) *Farm Focus: The 80s*, Canberra: National Farmers' Federation.

National Farmers' Federation (1986) *A Policy for Rural and Economic Growth*, Canberra: National Farmers' Federation.

National Farmers' Federation (1993) *New Horizons: A Strategy for Australia's Agifood Industries*, Canberra: National Farmers' Federation.

Neuman, W.L. (1997) *Social Research Methods: Qualitative and Quantitative Approaches* (3rd edn), Boston: Allyn and Bacon.

O'Malley, P. (1996) 'Indigenous governance', *Economy and Society*, **25**(3): 310-326.

O'Malley, P., Weir, L. and Shearing, C. (1997) 'Governmentality, criticism, politics', *Economy and Society*, **26**(4): 501-517.

Parliament of Queensland (1967a) *Farmers' Assistance (Debts Adjustment) Act*, Brisbane: Government Printer.

Parliament of Queensland (1967b) 'Farmers' Assistance (Debts Adjustment) Bill', *Queensland Legislative Assembly Debates*, (Vol. 245) Brisbane: Government Printer.

Parliament of Queensland (1970) 'Farmers' Assistance (Debts Adjustment) Act Amendment Bill', *Queensland Legislative Assembly Debates*, (Vol. 255) Brisbane: Government Printer.

Parliament of Queensland (1985) 'Queensland Industry Development Corporation Bill', *Debates, Queensland Legislative Assembly*, (Vol. 301) Brisbane: Government Printer.

Parliament of Queensland (1986) *Queensland Industry Development Corporation Act*, Brisbane: Government Printer.

Parliament of Queensland (1994a) *Queensland Rural Adjustment Authority Act*, Brisbane: Government Printer.

Parliament of Queensland (1994b) 'Rural Adjustment Authority Bill', *Debates, Queensland Legislative Assembly*, (Vol. 330) Brisbane: Government Printer.

Poiner, G. (1990) *The Good Old Rule: Gender and Other Power Relationships in a Rural Community*, Sydney: Sydney University Press.

Portela, J. (1994) 'Agriculture: is the *art de la localite* back? The role and function of indigenous knowledge in rural communities', in J.B. Dent and M.J. McGregor (eds), *Rural and Farming Systems Analysis: European Perspectives*, Wallingford, UK: CAB International: 269-279.

Pusey, M. (1991) *Economic Rationalism in Canberra: A Nation Building State Changes its Mind*, Sydney: Cambridge University Press.

Queensland Government (1992) *Drought: Managing for Self-Reliance*, Brisbane: Government Printer.

Redclift, M. (1993) 'Sustainable development: concepts, contradictions, and conflicts', in P. Allen (ed.), *Food for the Future. Conditions and Contradictions of Sustainability*, New York: John Wiley and Sons: 169-192.

Richardson, L. (1994) 'Writing: a method of inquiry', in N.K. Denzin and Y.S. Lincoln (eds), *Handbook of Qualitative Research*, Thousand Oaks, CA: Sage: 516-529.

Robertson, G. (1997) 'Managing the environment for profit', in *Outlook '97, Volume 2: Agriculture*, Proceedings of the National Agricultural Outlook Conference, Canberra, February 4-6: 75-79.

Rose, N. (1993) 'Government, authority and expertise in advanced liberalism', *Economy and Society*, **22**(3): 283-299.

Rose, N. (1996) 'The death of the social?: re-figuring the territory of government, *Economy and Society*, **25**(3): 327-356.

Rose, N. (1999) *Powers of Freedom: Reframing Political Thought*, Cambridge: Cambridge University Press.

Rose, N. and Miller, P. (1992) 'Political power beyond the state: problematics of government', *British Journal of Sociology*, **43**(2): 173-205.

Rural Adjustment Scheme Advisory Council (1996) *Annual Report 1995-96*, Canberra: Australian Government Publishing Service.

Rural Reconstruction Board of Queensland (1972) *Annual Report 1972*, Brisbane: Government Printer.

Rural Weekly (1999) 'Bitter row over exceptional circumstances', Article, May 7.

Schapper, H. (1974) 'Policy needs for farm adjustment', in C.D. Throsby (ed.), *Agricultural Policy: Selected Readings*, Ringwood: Penguin: 96-107.

Schmalkuche, P. (1986) *Cattlemen's Union: The First Decade*, Brisbane: Brolga.

Schultz, T.W. (1949) *Production and Welfare of Agriculture*, New York: Macmillan.

Scott, R., Coaldrake, P., Head, B. and Reynolds, P. (1986) 'Queensland', in B. Galligan (ed.), *Australian State Politics*, Melbourne: Longman Cheshire: 51-73.

Smyth, P. (1994) *Australian Social Policy: The Keynesian Chapter*, Sydney: UNSW Press.

Standen, B.J. and Musgrave, W.F. (1969) 'The agricultural adjustment problem with particular reference to Australia', in J.P. Makeham and J.G. Bird (eds), *Problems of Change in Australian Agriculture*, Armidale: Department of University Extension, University of New England: 13-30.

Stayner, R. (1994) *A Study of Farm Adjustment in the Bland Shire: Final Report to the NSW Rural Assistance Authority*, Sydney: New South Wales Rural Assistance Authority.

Stayner, R. (1996) 'Policy issues in farm adjustment', in Dept. of the Parliamentary Library (ed.), *Australian Rural Policy Papers 1990-95*, Canberra: Australian Government Publishing Service: 161-197.

Stehlik, D., Gray, I. and Lawrence, G. (1999) *Drought in the 1990s: Australian Farm Families' Experiences*, Canberra: Rural Industries Research and Development Corporation.

Stenson, K. (1998) 'Beyond histories of the present', *Economy and Society*, **27**(4): 333-352.

Stenson, K. and Watt, P. (1999) 'Governmentality and the death of the social?: a discourse analysis of local government texts in South-east England', *Urban Studies*, **36**(1): 189-201.

Stilwell, F. (1993) 'Economic rationalism: sound foundations for policy?' in S. Rees, G. Rodley and F. Stilwell (eds), *Beyond the Market: Alternatives to Economic Rationalism*, Sydney: Pluto Press: 27-37.

Strange, S. (1996) *The Retreat of the State: The Diffusion of Power in the World Economy*, Cambridge: Cambridge University Press.

Synapse Consulting (1992) *Report of the Review of the Rural Adjustment Scheme*, Canberra: Department of Primary Industries and Energy.

Taylor, S.J. and Bogdan, R. (1998) *Introduction to Qualitative Research Methods: A Guidebook and Resource* (3rd edn), New York: John Wiley and Sons.

Trebeck, D. (1990) 'Farmer organizations', in D.B. Williams (ed.), *Agriculture in the Australian Economy* (3rd edn), Sydney: Sydney University Press: 127-143.

Vanclay, F. (1992) 'The social context of farmers' adoption of environmentally sound farming practices', in G. Lawrence, F. Vanclay and B. Furze (eds), *Agriculture, Environment and Society: Contemporary Issues for Australia*, Melbourne: Macmillan: 94-121.

Vanclay, F. and Lawrence, F. (1995) *The Environmental Imperative: Eco-Social Concerns for Australian Agriculture*, Rockhampton: Central Queensland University Press.

Watson, A.S. (1979) 'Rural policies', in A. Patience and B. Head (eds), *From Whitlam to Fraser: Reform and Reaction in Australian Politics*, Melbourne: Oxford University Press: 157-172.

Watson, C. (1992) 'An ecologically unsustainable agriculture', in G. Lawrence, F. Vanclay and B. Furze (eds), *Agriculture, Environment and Society: Contemporary Issues for Australia*, Melbourne: Macmillan: 19-32.

Weiss, L. (1997) 'Globalization and the myth of the powerless state', *New Left Review*, **225**: 3-27.

White, D.H. and Bordas, V.M. (eds) (1996) *Indicators of Drought Exceptional Circumstances*, Canberra: Bureau of Rural Sciences.

Yeatman, A. (1994) *Postmodern Revisionings of the Political*, New York: Routledge.

APPENDIX A: DOCUMENTS ANALYZED IN THE STUDY

As part of the research for this book various historical documents were collected that recorded discourses on the governing of agricultural change between 1967 and 1997. These documents are outlined below and divided into policy documents, official reviews/reports, parliamentary debates, and also a selected list of agricultural economics articles. The list below includes only those documents used to reconstruct agricultural change as a governable problem. Other literature is used also in the course of the book to help frame and analyze these documents.

POLICY DOCUMENTS AND ACTS OF PARLIAMENT

Commonwealth:

1997 – Department of Primary Industries and Energy *Agriculture - Advancing Australia*
1992 – Commonwealth of Australia *Rural Adjustment Act*
1992 – Commonwealth of Australia *National Drought Policy: Managing for Risk and Productivity*
1991 – Dept. of the Prime Minister and Cabinet *Building a Competitive Australia*
1988 – Hon. Paul Keating (Treasurer) *May Economic Statement*
1988 – Hon. John Kerin (Minister for Primary Industries) Senator P. Cook (Minister for Resources) *Primary Industries and Resources: Policies for Growth*
1988 – Commonwealth of Australia *States and Northern Territory Grants (Rural Adjustment) Act*
1986 – Federal Treasurer, Hon. Paul Keating's 'Banana Republic' Statement
1986 – Commonwealth of Australia *Economic and Rural Policy: A Government Policy Statement*

1988 – Commonwealth of Australia *States and Northern Territory Grants (Rural Adjustment) Act*
1983 – Hon. John Kerin (Minister for Primary Industry) *Rural and Regional Policy*
1976 – Commonwealth of Australia *State Grants (Rural Adjustment) Act*
1973 – Commonwealth of Australia *Industries Assistance Commission Act*
1971 – Commonwealth of Australia *States Grants (Rural Reconstruction) Act*
1970 – Commonwealth of Australia *Marginal Dairy Farms Agreement Act*

State:

1994 – Queensland Legislative Assembly *Queensland Rural Adjustment Authority Act*
1992 – Queensland Government *Drought: Managing for Self-Reliance*
1986 – Parliament of Queensland *Queensland Industry Development Corporation Act*
1976 – Parliament of Queensland *Rural Adjustment Agreement Act*
1970 – Parliament of Queensland *Farmer's Assistance (Debts Adjustment) Amendment Act*
1967 – Queensland Legislative Assembly *Farmer's Assistance (Debts Adjustment) Act*

Farm Organizations:

1993 – National Farmers' Federation *New Horizons: A Strategy for Australia's Agrifood Industries*
1986 – National Farmers' Federation *A Policy for Rural and Economic Growth*
1981 – National Farmers' Federation *Farm Focus: The 80s*

REVIEWS/REPORTS

Commonwealth:

1997 – J. McColl, R. Donald and C. Shearer *Rural Adjustment: Managing Change - Mid-term Review of the RuralAdjustment Scheme*
1996 – Department of Primary Industries and Energy *National Rural Finance Summit: Report of the Activating Committee*
1996 – Commonwealth of Australia *Report of the Drought Policy Review Task Force*
1995 – Commonwealth of Australia *Managing for the Future: Report of the Land Management Task Force*
1994 – Commonwealth of Australia *Rural Adjustment, Rural Debt and Rural Reconstruction: Report of the Senate Rural and Regional Affairs and Transport References Committee*
1992 – Commonwealth of Australia *National Drought Policy - Appropriate Government Responses to the Recommendations of the Drought Policy Review Task Force: First Report From the Senate Standing Committee on Rural and RegionalAffairs*
1992 – Synapse Consulting *Report of the Review of the Rural Adjustment Scheme*
1990 – Commonwealth of Australia *Report of the Drought Policy Review Task Force*
1988 – Coopers and Lybrand WD Scott *Report on the Review of the Rural Adjustment Scheme*
1984 – Industries Assistance Commission *Report on Rural Adjustment*
1982 – J. Balderstone, L. Duthie, F. Jarrett, D. Eckersley and J. McColl *Agricultural Policy: Issues and Options for the 1980s. Working GroupReport to the Minister for Primary Industry* (Green Paper)
1976 – Industries Assistance Commission *Rural Reconstruction*
1974 – S. Harris, J. Crawford, F. Gruen and N. Honan *Rural Policy in Australia. A Working Group Report* (Green Paper)
1970 – Bureau of Agricultural Economics *Debt Reconstruction and Farm Adjustment: The Immediate and Longer Term Needs for Debt Reconstruction and Farm Adjustment with Special Reference to the Sheep Industry*

PARLIAMENTARY DEBATES

Commonwealth:

1992 – House of Representatives *Rural Adjustment Bill*
1988 – House of Representatives *States and Northern Territory Grants (Rural Adjustment) Bill*
1985 – House of Representatives *States and Northern Territory Grants (Rural Adjustment) Bill*
1976 – House of Representatives *States Grants (Rural Adjustment) Bill*
1973 – House of Representatives *Industries Assistance Commission Bill*
1971 – House of Representatives *States Grants (Rural Reconstruction) Bill*
1970 – House of Representatives *Rural Reconstruction: Discussion of Matter of Public Importance*
1970 – House of Representatives *Australian Rural Economy: Discussion of Matter of Public Importance*
1970 – House of Representatives *Marginal Dairy Farms Agreement Bill*

State:

1994 – Parliament of Queensland *Rural Adjustment Authority Bill*
1985 – Parliament of Queensland *Queensland Industry Development Corporation Bill*
1970 – Parliament of Queensland *Farmer's Assistance (Debts Adjustment) Amendment Bill*
1967 – Parliament of Queensland *Farmer's Assistance (Debts Adjustment) Bill*

SELECTED ARTICLES BY AGRICULTURAL ECONOMISTS

1970 – S. Harris *Farm Adjustment and the Role of Government*
1969 – B. Standen and W. Musgrave *The Agricultural Adjustment Problem with Particular Reference to Australia*
1967 – D. McKay *The Small Farm Problem in Australia*

APPENDIX B: INTERVIEWING AND SAMPLING PROCEDURES

The interviewing component of this research project commenced in June 1997 and was completed in July 1998, involving semi-structured interviews. An outline of who was spoken to, on which dates, and in what location interviews took place, is presented below in tabular form.

Participants	Date of Interview	Location
Commonwealth Public Servants (4 Males, 1 Female)	June 30-July 1 1997	Canberra
Queensland State Public Servants (2 Males, 1 Female)	September 29-30 1997	Brisbane
Queensland State Public Servant (1 Male)	April 17 1998	Brisbane
Farm Organization Representative (1 Male)	April 20 1998	Brisbane
Farm Organization Representatives (3 Males)	May 14 and 17 1998	Rockhampton
Commonwealth Public Servants (2 Males)	July 6 1998	Canberra
Farm Organization Representative (1 Male)	July 7 1998	Canberra
Commonwealth Public Servants – Member Check (2 Males, 1 Female)	July 7 1998	Canberra

Sixteen people in total were interviewed. Of these, seven were Commonwealth public servants (Department of Primary Industries and Energy, Rural Industries Research and Development Corporation), four were Queensland State public servants (Queensland Department of Primary Industries, Queensland Treasury, Office of Rural Communities, Queensland Rural Adjustment Authority) and five were representatives of farm organizations (Queensland Graingrowers, United Graziers Association, Cattlemen's Union – these three organisations have now amalgamated under the banner of AgForce). It is important to note that the interviews on July 7 1998 were 'member-checking' interviews (see Burns, 1994; Merriam, 1998). The member-checking interviews provided an excellent opportunity to help *crystallize* (Richardson, 1994) the findings by questioning Commonwealth public servants on issues that had been raised in discussions with other participants. This allowed further comment, exploration and reflection on those issues that were raised by participants in the first round of interviews..

Participants for the interviews were chosen through *purposive* (Neuman, 1997) or *theoretical* (Lincoln and Guba, 1985; Taylor and Bogdan, 1998) sampling using three techniques: existing contacts from previous research, individuals suggested by initial participants, and people who were found to have some involvement with the Rural Adjustment Scheme (RAS). Purposive sampling selects usually cases with a specific purpose in mind. It is appropriate in three situations: first, a researcher uses it to select cases that are particularly informative; second, he or she may use it to select members of a difficult-to-reach population; and third, one might use it to identify particular types of cases for in-depth investigation (Neuman, 1997:206). In this project the first situation was most applicable as those participants involved directly with the making of policy on farm adjustment were assumed to have the most knowledge on the topic, and play a key role in the management of agricultural change.

Prospective interview participants were contacted initially by phone. As contact had been established with some of these people in a previous study, it was believed that a phone call would be the most suitable way of reintroduction and gaining the rapport that Glesne and Peshkin (1992) and Fontana and Frey (1994) note is so important in qualitative research. This strategy turned out to be highly successful with many participants showing an interest in the research and even suggesting other possible informants. It improved greatly the finding of suitable participants, and many of the people suggested were eventually interviewed.

The interviews followed a conventional semi-structured approach with no fixed wording or ordering of questions (Minichiello et al., 1990:92). Interviews were recorded on audiotape and most interviews were between 40 and 60 minutes in length. Textual material from the interview transcripts was coded by the author with data selected and categorized based on a coding scheme constructed from the research objectives and interview themes.

APPENDIX C: INTERVIEW DISCUSSION TOPICS

- Can you tell me a little about your involvement with the RAS?
- Why do you believe the RAS was originally created?
- Has the RAS assisted the farmers it was designed to assist?
- To what extent was the RAS a welfare program?
- To what extent was the Federal Government helped or hindered by State Governments in administering the RAS?
- How important have farm lobby groups been in terms of maintaining RAS assistance?
- Why have successive governments not left rural adjustment issues to the market alone? (ie. why has there been intervention at all?)
- What do you think are the reasons why the RAS has become focused in recent years on farmer self-reliance?
- Do you think that self-reliance alone can address the economic, environmental and social problems facing farmers? Why or why not?
- Do you think that the Commonwealth has some obligation to assist farm families in certain circumstances? Examples?
- How do you see the RAS as having affected rural Australia in general terms?
- What do you see as the future for rural adjustment now that the RAS has largely been restructured?

Additional questions for interviews with State public servants and representatives of farm organizations:

- Has the RAS been a difficult policy to interpret? Why or why not?
- How influential do you think that your department/organization has been in terms of maintaining RAS assistance?
- Do you think that Commonwealth objectives have been followed closely by State administrations?
- What do you think are the reasons why the RAS has become more focused on farmer self-reliance and management over the years? Has your department/organization supported or opposed this shift?

Discussion topics for Member Checking Interviews conducted with Commonwealth public servants: July 6-8 1998.

- What review of the RAS do you think marked the most decisive shift towards a more pro-active, long-term and self-reliant approach to rural adjustment? Why?
- Do you believe that RAS 1992 separated business and welfare objectives clearly enough? If not, why?
- Do you think that farmers can ever be expected to become totally self-reliant? If no, what are some of the main limitations?
- Should government policies continue to recognize that there are some events that a farmer cannot be reasonably expected to plan for such as severe drought and commodity price collapses?
- Do you think that the recently-released 'Triple A' package provides the appropriate market signals to farmers in terms of risk management and productivity? Do you think that the RAS ever provided the 'right' market signals?
- Farm organizations believe that they have, overall, been very successful in lobbying to maintain RAS assistance (ie. interest subsidies) to farmers. What are your opinions on this issue?
- Farm organizations are generally supportive of the shift towards greater farmer self-reliance. Do you agree that these organizations have supported Commonwealth objectives in this respect?
- Farm organizations believe they are representing all farmers/graziers in terms of promoting a more self-reliant approach. What are your opinions on this issue?
- Do you think that the Commonwealth has taken a flexible approach to the State administration of the RAS?
- Which areas of RAS have been subject to the most flexible interpretation by the states?
- Has Queensland's approach to rural adjustment been consistent with Commonwealth objectives of productivity and self-reliance? Why or why not?
- State public servants and farm organizations believe that the drought in Queensland has hindered farmers there becoming independent of government support. What are your opinions on this issue?
- Do you believe that the Departments of Treasury and Finance have played an increasingly prominent role in the formulation and evaluation of rural adjustment policy? If so, when did they start becoming influential?
- What do you see as the future for Commonwealth involvement in rural adjustment? Do you think that we will see State governments assuming greater responsibility for rural adjustment and assistance schemes in the future?

INDEX

Q

R

S

T

U

W